Frontier Military Series
XXXIII

BEFORE CUSTER

Surveying the Yellowstone 1872

Edited by
M. John Lubetkin

THE ARTHUR H. CLARK COMPANY
An imprint of the University of Oklahoma Press
Norman, Oklahoma
2015

Also by M. John Lubetkin

Jay Cooke's Gamble: The Northern Pacific Railroad, the Sioux, and the Panic of 1873
Custer and the 1873 Yellowstone Survey: A Documentary History
Custer's Gold
Union College's Class of 1868: The Unique Experiences of Some "Average" Americans

EPIGRAPH (P. VI) SOURCE

Northern Pacific Railway, Annual Reports, 1870–1968, submitted to the Department of the Interior, Annual Report for 1872, September 30, 1872, pp. 12–13, Minnesota Historical Society, St. Paul (microfilm M-369).

Library of Congress Cataloging-in-Publication Data

Before Custer : surveying the Yellowstone, 1872 / edited by M. John Lubetkin.
pages cm. — (Frontier military series ; volume 33)
Includes bibliographical references and index.
ISBN 978-0-87062-431-5 (hardcopy : alk. paper) 1. Yellowstone River Valley—History—19th century—Sources. 2. Yellowstone River Valley—Surveys—History—19th century—Sources. 3. Railroads—Surveying—History—19th century—Sources. 4. Northern Pacific Railroad Company—History—19th century—Sources. 5. Indians of North America—Wars—1866–1895—Sources. 6. Dakota Indians—Wars—Sources. I. Lubetkin, M. John. II. Title: Surveying the Yellowstone, 1872.
F737.Y4B44 2015
978.6'3—dc23

2014035476

Before Custer: Surveying the Yellowstone, 1872 is Volume 33 in the Frontier Military Series.

The paper in this book meets the guidelines for permanence and durability of the Committee on Production Guidelines for Book Longevity of the Council on Library Resources, Inc. ∞

1 2 3 4 5 6 7 8 9 10

To my sister,
Elizabeth Lubetkin-Lipton,
and my brother,
James Green Lubetkin

I exceedingly regret to have to report to the Department of the Interior that our [surveyors] and their [army] escort have been resisted and fought by armed forces of Indians, in such numbers and with such spirit, as to stop our work. This company's engineers have in effect been driven by the Indians out of the Yellowstone Valley . . . and we have been forced to suspend [their work] for the season between the Missouri River and the Rocky Mountains.

The sincere intention of the Government to protect this public work was defeated by superior force. It is cause for both mortification and regret that an enterprise so wholly national and beneficent as is the construction of the Northern Pacific Railroad should be thus interrupted and held back.

NORTHERN PACIFIC RAILWAY
ANNUAL REPORT
SEPTEMBER 30, 1872

Contents

Appendices

Illustrations

Maps

Introduction

Before Custer is told by white male participants in diaries, letters, reports, and newspaper articles from previously unpublished documents. This is the story of two strategic Indian victories against army forces consisting of almost a thousand men. The consequences of these U.S. defeats all but destroyed the nation's largest and (supposedly) safest private banking house, brought Sitting Bull to national attention, directly led to George Armstrong Custer's transfer to Dakota, and forever weakened Northern Pacific Railway plans to dominate a half-dozen northwestern states. This is also the story of business corruption, alcoholic army commanders, and the Northern Plains Indians' strengths and weaknesses during their last years of semi-independence.

The real Northern Pacific story is not that of the expanding white settlement pushing against Indian lands, nor is it of a "rush" for mineral wealth. Rather, these events occurred because one man believed that, as he was "God's chosen Instrument," it was his mission to build the railroad.[1]

I

If Jay Cooke is remembered at all today, it is as a robber baron with a Civil War connection. However, when the war ended, Cooke was a genuine hero to the (northern) American public, standing taller in his day than J. P. Morgan. Exceptionally intelligent, invariably honest in handling public money, and a prototype humanitarian, he nevertheless

[1]See Lubetkin, *Jay Cooke's Gamble,* chap. 1, "God's Chosen Instrument."

bribed, threatened, and bulldozed congressmen as few (if any) have ever done. He made his fortune trading on insider information, violated virtually every present-day antitrust concept, and owned five homes, including one that for a time was the largest private residence in the United States, and another built on his own Lake Erie island.

When the Civil War began, every literate American knew of George Washington's problems in feeding, clothing, sheltering, arming, and especially paying his troops. He once faced down a mob of unpaid soldiers. By mid-1862, as the war showed no signs of abating, both sides—with good reason—worried whether they could continue to finance the conflict. While the South never satisfactorily answered this question, the North and Abraham Lincoln's financing problems were met by forty-year-old Cooke.

Over a three-year period Cooke sold or directed the sale of some $1.6 billion in war bonds, equaling more than 25 percent of every Northern dollar raised.[2] This "tumbling river of cash," as Carl Sandburg called it—in tens of thousands of small payments—was *never* tainted by scandal.[3] The public venerated Cooke for handling its money honestly, and no Union general faced the scarcities that had concerned Washington. By 1865, four years after launching his own bank in a borrowed Philadelphia office, Cooke was America's leading banker.

Yet by 1870 Cooke was under siege by those who had business interests in the just-completed Union Pacific–Central Pacific transcontinental railroad. The cause was his 1869 decision to finance the Northern Pacific Railway. Soon the transcontinental backers systematically slowed and undermined Cooke and the Northern Pacific at every opportunity. Congress blocked Cooke's attempts to get federal dollar subsidies similar to the transcontinental road, forcing him to "settle" for 49 million acres in land grants—larger than all New England.[4]

Additionally, the British and Canadian governments feared—with considerable justification—that western Canada, then being overrun by

[2]"More than 25 percent" is a conservative figure. Cooke sales cleared the psychological way for others to sell bonds, and he assisted and/or advised other bond sellers, including individual states.

[3]Sandburg, *Abraham Lincoln,* 2:608.

[4]The 49 million acres is an order-of-magnitude estimate. Some estimates are higher, but Congress's grant overlapped earlier ones (especially in Minnesota), leaving the actual number in doubt.

Americans, would soon become more stars in the American flag. Thus, when Cooke looked to Europe for financing, the British, working behind the scenes, thwarted him at every turn.

In the United States, Cooke's weak bond sales were matched by abysmal construction. Although Cooke and Northern Pacific president J. Gregory Smith, who was also president of the ninth largest U.S. railroad, the Vermont Central, agreed to move forward in late 1869, Smith failed to hire construction companies until mid-1870. After a sham groundbreaking ceremony in February, real work didn't begin until August. Almost immediately, crossing Minnesota's wetlands/lake country proved more difficult than anticipated.

In November Smith ordered work to continue through the winter, despite Minnesota's legendary cold. Men died by the score, and amputations resulting from exposure and injuries were in the hundreds. When the spring thaw arrived, most of the railway buckled or sank, and much of 1871 had to be devoted to rebuilding while siphoning off crews from other projects. Worse, beset by Vermont Central financial problems, Smith seldom left New England, visiting Minnesota only five times in four years.

Could anything else go wrong? Of course. Although Smith was tireless and intelligent, he was also fundamentally dishonest. He himself likely profited and certainly kept his Vermont Central afloat with Crédit Mobilier–type contracts for the Northern Pacific, all possible from Cooke's bond sales. As one writer summarized Smith's management: "He soon came under attack for the same sort of financial tactics he had perfected in Vermont, as a growing number of observers charged him with extravagance and misuse of NP funds to enrich himself and his associates. In the face of this growing criticism, and with the NP close to bankruptcy, Smith resigned."[5] What seems clear is that Smith appears to have had no idea his parasitic policies could bring down Cooke's financial empire, always believing that Cooke's warnings were mere bluffs. Ironically, Smith was no Johnny-come-lately to the northern transcontinental cause, having believed passionately in such a railroad since the early 1850s.

---

[5]"John Gregory Smith," entry by J. Kevin Graffagnino, in *Encyclopedia of American Business, History, and Biography: Railroads in the Nineteenth Century*, 366.

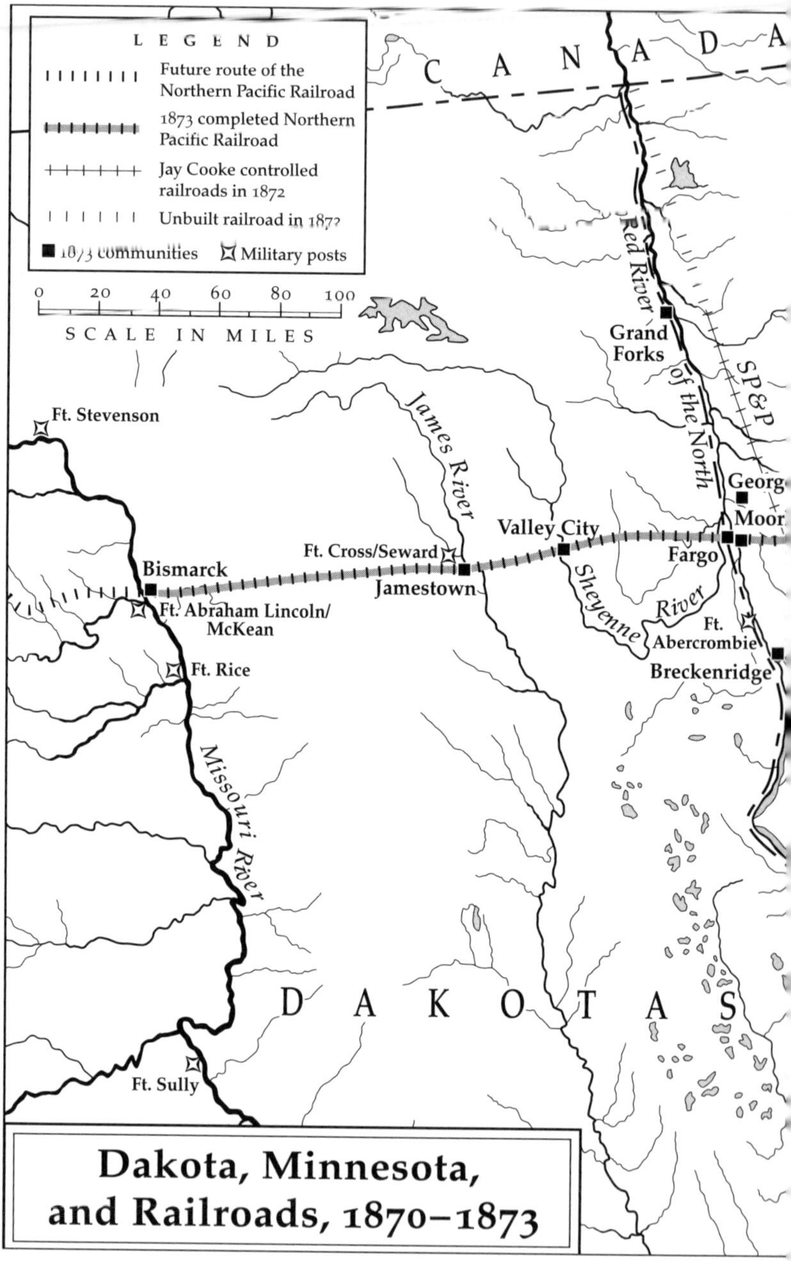

Dakota, Minnesota, and Railroads, 1870–1873

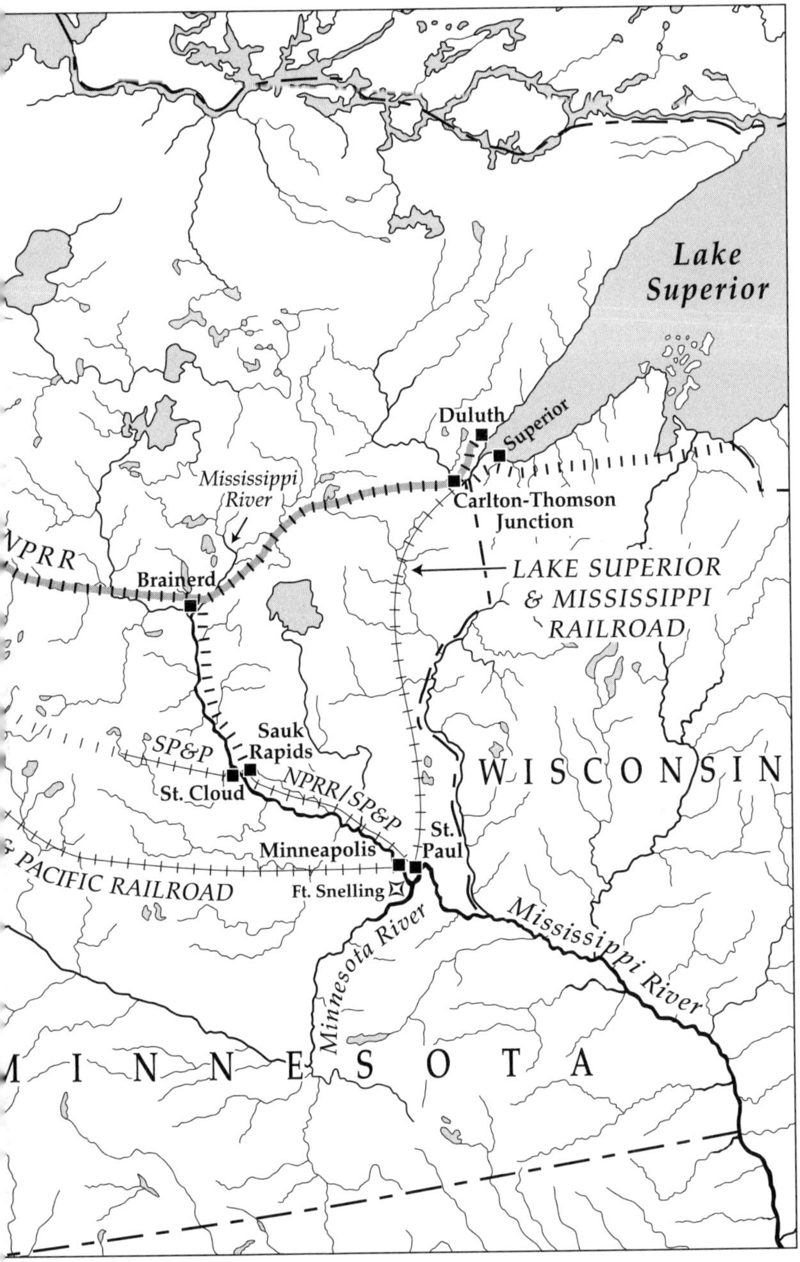

Map by Vicki Trego Hill, adapted from an 1883 Northern Pacific Railroad map in the author's collection. *Copyright © 2015 The University of Oklahoma Press.*

The Northern Pacific had planned to reach the Missouri in midsummer 1872. However, the Red River wasn't bridged at Fargo-Moorhead until July, leaving Dakota construction far behind schedule. Next, filings for government certification and public sale of *millions* of Minnesota acres were delayed by Smith, ending chances of 1872 income from this source. As pressure built, hundreds if not thousands of workers were shifted willy-nilly from one project to the next, resulting in expensive unneeded or unfinished routes.

If this was no way to run a railroad, in truth railroad management was barely out of its infancy. Although huge technical strides were made during the war, management practices and especially staff changed little in forty years.[6] The barely literate Cornelius Vanderbilt, for example, supposedly oversaw his business empire from a stand-up desk with clerks reading him messages to which he dictated responses. Other than the construction side, which required skilled engineers, Smith's staff support, like Vanderbilt's, was limited. Even had he been totally honest, the problems arising from constructing two thousand miles of track *plus* managing the struggling Vermont Central were simply too Herculean.

In May 1872, Cooke quietly forced Smith's resignation, but amazingly, although he had invested so much political capital, he had no replacement ready to step in! So Smith remained de facto president, maneuvering to get back into office. Not until September, when his Vermont Central collapsed, did Smith walk away from the Northern Pacific. Yet earlier, Smith awarded the key Fargo to Bismarck construction contract to a group headed by Dr. Walter A. Burleigh, a physician, attorney, and disreputable Indian agent devoid of railroad experience. In classic Crédit Mobilier fashion, Burleigh then sold the contract to others. No one knows the details, but it is naïve to think Smith was not in some way compensated.

* * *

From the beginning, Cooke's bank picked up the difference between Northern Pacific bond sales and the railroad's expenditures. In Europe

---

[6]The first experimental steam-powered locomotives (versus horse-drawn or "gravity" trains) were tried out in Baltimore and Charleston, South Carolina, circa 1830. One earlier locomotive ran on a short, circular track in New Jersey. The first regularly scheduled steam railroad was the Mohawk & Hudson, which began operations between Albany and Schenectady, New York, in August 1831.

the 1870 Franco-Prussian War froze much European capital, and Great Britain's hostility made sales insignificant. While bond sales were Cooke's forte in the United States, over six thousand miles of track was being constructed annually in more economically developed areas and under more secure conditions. Thus much of the money Cooke had anticipated was spent or committed to safer projects.

Cooke's bank could pick up only so much of the slack, and by June 1872 the wolf was at his door. Shortly before both surveys began, a younger partner, Harris C. Fahnestock, detailed the company's financial problems, and two months later Cooke, visibly embarrassed, told aghast Northern Pacific directors that his bank was no longer able to advance funds: it was almost out of cash.

To the extent the public focused on the Northern Pacific, it was on the 1872 surveyors and their army escorts who were again to enter the Sioux-controlled Yellowstone valley. However, given the military's support, how could serious problems be anticipated?

In 1872 the army totaled 31,450 officers and men, their service determined by function or geographic requirements.[7] Lt. Gen. Philip H. Sheridan commanded the nation's largest region, the 6,771-man Division of the Missouri, which stretched from Mexico to Canada. The largest subdivision was the Department of Dakota, with 3,719 officers and men posted in Minnesota and Territorial Dakota and Montana. In turn, that was divided into three districts and Fort Buford (an independent command), which encompassed some 370,000 square miles. The largest of these, with 1,753 men, was the Dakota or Middle District with troops at numerous forts and Indian agencies (reservations).

Before Custer takes place entirely within the Department of Dakota, in 1872 commanded by Maj. Gen. Winfield Scott Hancock. Hancock, as every schoolboy knew one hundred years ago, was Gettysburg's hero, having turned back Pickett despite suffering a grievous abdominal wound. A West Point classmate of Grant's, Hancock was an active Democrat who made a strong showing at the 1868 presidential convention, inflating his already oversized ego. Detesting both Grant and Sheridan, he felt his posting to St. Paul was personally and politically motivated.

Joint Northern Pacific–U.S. Army plans called for survey teams leaving from Fort Ellis, adjacent to Bozeman, Montana, and Fort Rice

[7]U.S. Department of War, Annual Report, November 1872, 35–38.

on the Missouri, twenty miles south of present-day Bismarck, North Dakota. The two surveying expeditions (as they were often called in contemporary newspapers) were to meet where the Powder River enters the Yellowstone, thus completing the railroad's two thousand mile "line of survey" between Lake Superior and Puget Sound.

Both surveys had important secondary missions. The 1871 eastern survey was from Bismarck to where Glendive Creek flowed into the Yellowstone. In 1872, the railroad hoped to find a more direct route from Bismarck to the Powder River, using a shortcut eliminating at least fifty miles and $750,000 of track. The western survey, on its return, was to leave the Yellowstone near present-day Forsyth, cut north to the Musselshell River, and follow it west toward Helena, thus avoiding the difficult and expensive Bozeman Pass. Northern Pacific management fully recognized the economic implications of both routes, but common sense dictated possessing all engineering options before making final decisions.

II

Entering 1872, neither army nor railroad executives thought that a small band of "militant" Indians could stand in their way.[8] After all, the unstated policy of letting Northern Plains Indians kill off their remaining buffalo (and themselves) couldn't be going more effectively. Buffalo had basically vanished in Dakota, and the overwhelming majority of the Teton or Lakota Sioux had no choice but to live on agencies during the winter. Moreover, Sioux hunting land was increasingly contested by white hunters, Crows, Flatheads, Nez Perce, Blackfeet, and Métis ("mixed bloods") from northeastern Dakota and adjacent Canada. Worse, the Sioux were trapped in a Catch-22 of their own making: to purchase their arms and ammunition, the most valuable buffalo trading hide was a three-year-old female's—an arithmetical recipe for ecological disaster. In retrospect, it seems fair to assume that, regardless of the Northern Pacific's existence, sustainable numbers of buffalo would have vanished in less than a decade.

[8]Indians were referred to by the surveyors in numerous ways: as the *Indian problem,* as *savages,* or as *Mr. Lo* (as in "Lo the poor Indian"). Anyone hunting for the term "redskin" will not be disappointed; although its use was not infrequent, it was not used in a pejorative manner.

Under Sitting Bull's dynamic spiritual and political leadership, some five hundred lodges (with two or three warriors per lodge) attempted to retain their traditional way of life. His year-round followers represented perhaps 15 percent of the Lakota (Blackfeet, Brulé, his own Hunkpapa, Minneconjous, Oglala, San Arcs, and Two Kettle), as well as elements of Arapaho, Kiowa, and Northern Cheyenne, and survivors from the decimated Dakota (Santee) and Nakota (Waciyena) Sioux who had lived in Minnesota and Iowa. These Indians now wintered in northeast Montana, but were not so proud as to reject government assistance.

At first, Sitting Bull had not been aware of the 1871 surveys. Only in October did he learn their purpose: nothing less than building a railroad through the Yellowstone valley. Eastern Montana was his last hunting ground, and if he did not stop the Northern Pacific surveyors, his way of life would soon end. During the winter of 1871–72, word of the railroad's plans spread across the plains. By summer, Sitting Bull would be augmented by some 1,500 lodges who wintered at Dakota and Montana agencies.[9] One thing that Sitting Bull did not have to do was convince other Indians of the danger of railroads. Arapaho, Cheyenne, and Brulé Sioux living closest to the Platte River had experienced a series of crushing defeats, and with the completion of the transcontinental railroad, the central buffalo herds had vanished.

Sitting Bull, in his late thirties and with his back to the wall, was grimly determined to block the Northern Pacific. He had participated in battles against army and frontier militia troops at Dead Buffalo Lake (1863) and Killdeer Mountain (1864), both where the Sioux were pounded by artillery. Other than by ambush, the Sioux had invariably been bested. Sitting Bull was under no illusions, but he and his followers felt they had no choice.

Sitting Bull did not want war in 1872, and in a last effort to avoid it, he sent an emissary to the army to confirm the rumors. *Before Custer* begins in late March 1872 when Spotted Eagle visited Colonel David Stanley at Fort Sully.

---

[9]There are many sources, few of which agree with each other. Over the years I have found Robert M. Utley's estimates to be as objective and sensible as anyone's. Thus, for the number of Indians who gathered in 1872, the two thousand *lodges* is from Utley's *Lance and the Shield*, 107.

Montana and Dakota, 1864–1873

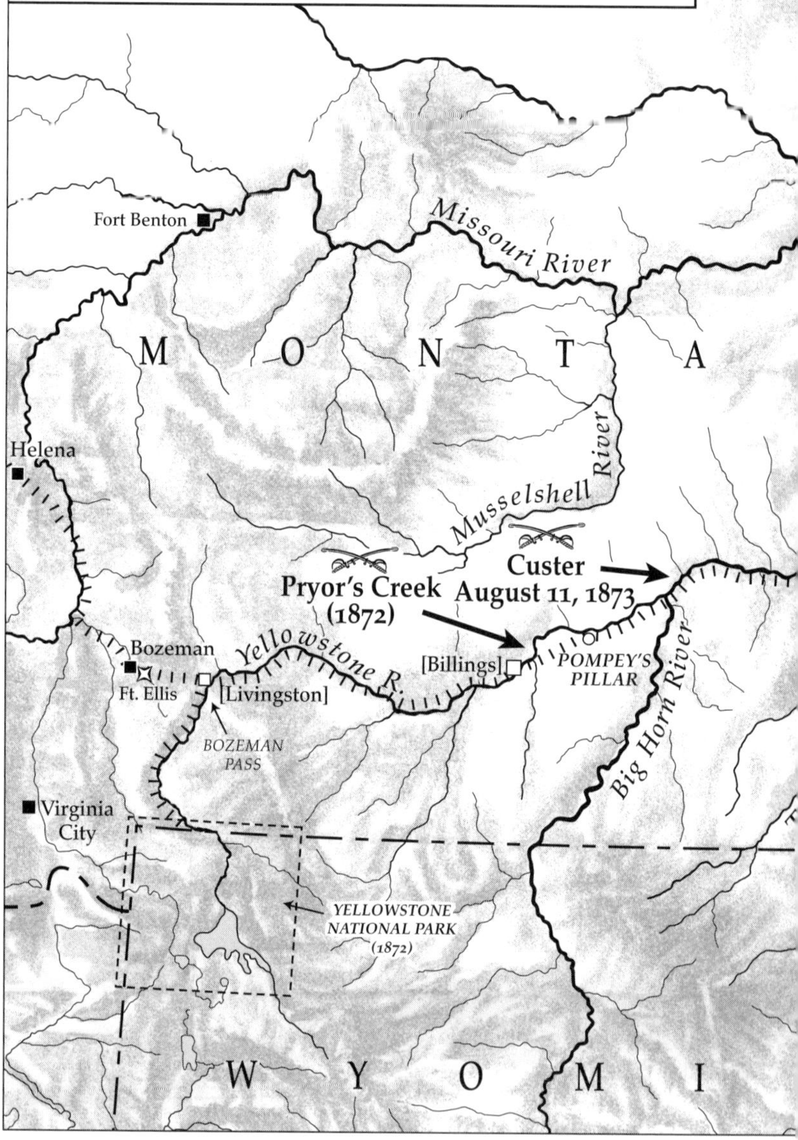

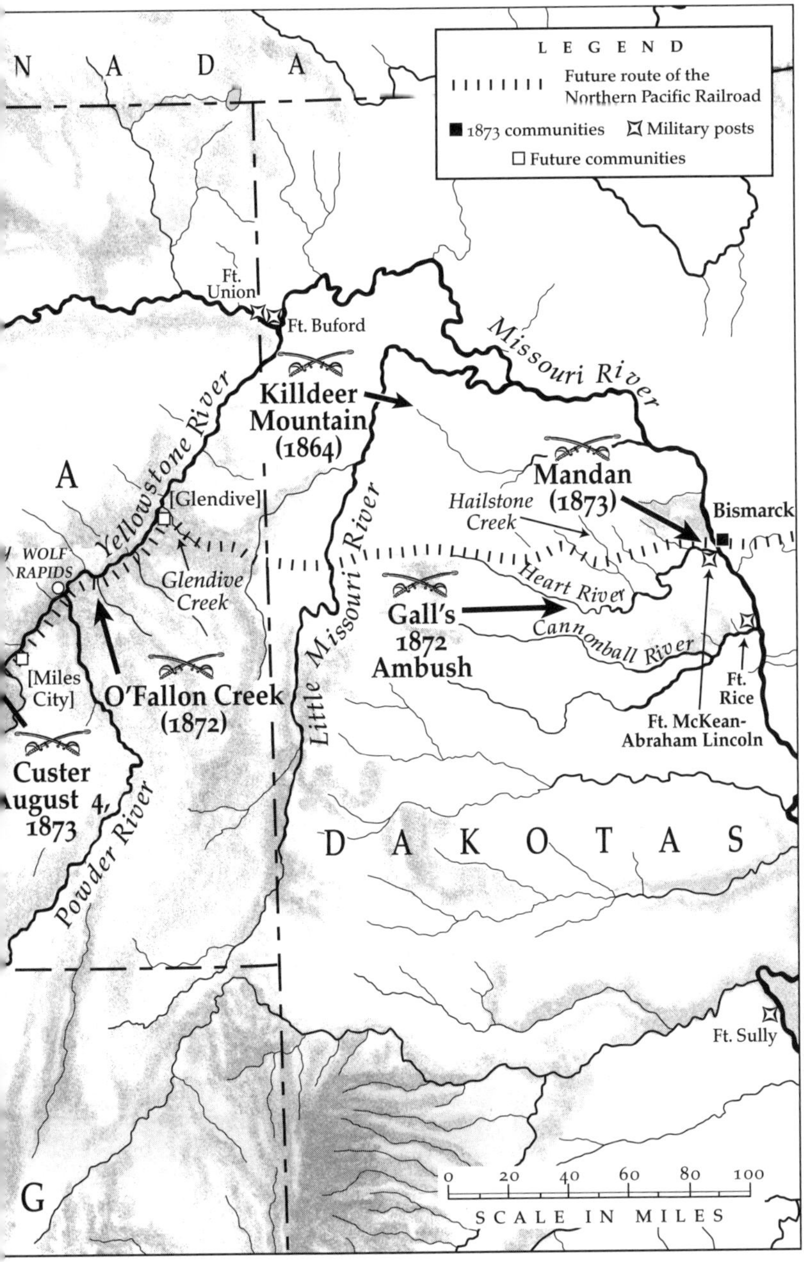

Map by Vicki Trego Hill, adapted from an 1883 Northern Pacific Railroad map in the author's collection. *Copyright © 2015 The University of Oklahoma Press.*

III

A word about primary sources. In general, western historians have ignored the four 1871 and '72 Yellowstone surveying expeditions. Some attention has been paid to the fifth (1873) survey for the simple reason that it included George Armstrong Custer, but many of these books act as if he had stepped into a vacuum. More often than not, the 1871–72 surveys have received only cursory treatment and then as a run-up to Custer.[10]

The largest depository of Northern Pacific materials is held by the Minnesota Historic Society. However, given the Northern Pacific's state of turmoil in late 1872, perhaps what is surprising is that anything survived. We know that the Northern Pacific's Helena office manager saved his letter-press book, but this five hundred page treasure trove is fast becoming illegible.[11] The few Northern Pacific documents and formal army reports are reproduced in their entirety.[12] Perhaps the most intriguing material is the report to Congress written by Maj. John Barlow of Sheridan's staff. Initially Barlow's report was so candid that Sheridan forced him to rewrite it. Sadly, even this modified document appears to have wrecked Barlow's career.

The only document published in the twentieth century is Lt. James H. Bradley's exceptionally candid posthumous description. Because he was killed fighting the Nez Perce, he never had the opportunity to edit out—as gentlemen did those days—his damning description of Baker's inebriated leadership.[13]

Contemporary newspapers are a mother lode of source material, although many have been lost, including a third of the issues of the

---

[10]See Bray, *Crazy Horse*; Larson, *Gall*; Wert, *Custer*; and Utley, *Lance and the Shield*.

[11]The letter-press process permitted ink to soak through from the first to a second layer of paper, leaving a blurred but readable copy *at the time*. Over the course of decades the ink has minutely spread, blurring words and leaving even enlarged xerographic copies often illegible.

[12]One report not included is that of Capt. Sanford C. Kellogg, Sheridan's official observer, whose short, dull, less than candid, politically correct submission is "Report to General Sheridan Concerning Stanley's 1872 Yellowstone Expedition."

[13]One document is Capt. Clarence E. Bennett's August 26 letter, the basis for Robertson's "'We Are Going to Have a Big Sioux War.'" While Bennett's letter is not printed in full, enough is available in the article or repeated in other sources so that it is not reproduced here.

Sioux City *Daily Journal.* However, frontier newspapers are classic examples of why today's editor must tread gingerly. They are often factually inaccurate—one story speaks of twenty thousand Sioux warriors—but marvelous for re-creating mood and feeling. Better yet, army personnel freely sent letters to the editor to publicly air their gripes and grudges, to say nothing of earning needed spare change.

Many primary sources are known to be missing. These include Northern Pacific western surveyor John Haydon's report and all western survey field diaries. Eastern documents include Maj. Robert Crofton's report describing, in what was likely a classic CYA document, the period in which Lt. Eban Crosby was killed.

Not to be found are Native American primary sources. The majority of the Indians who attempted to block the surveys spoke no English and certainly didn't write it. Only in the 1880s, as Indian survivors "adjusted" to reservation life, did their stories begin to unfold. However, inquiries by whites invariably focused on Custer and the Little Big Horn, and over the years little to no attention was paid by white interviewers to the 1872 Yellowstone surveys. Perhaps this is the reason so little has been written in the first place!

Racial prejudices played their role. Most officers held Indians in a similar low regard as they did African Americans. It never dawned on West Point trained officers that Sitting Bull's *military intelligence* could possibly be superior to theirs. In this writer's opinion, Indian leaders were more than capable of distinguishing between the military capabilities of each survey, and they went after the weakest target first. That their attack against Baker proved unsuccessful was a lack of discipline—the military bugaboo of Indians over three centuries.

No women accompanied the 1872 surveys, although the wife of Fort Rice's doctor, Linda Slaughter, briefly described the return of the surveyors. However, her account is that of anyone's aging faculties; memory begins to play tricks when trying to marry an incident with the date. Ultimately, like others, her remembrances were too short and uneven to be utilized.[14]

On editing: I have followed generally accepted modern spellings for Indian tribes. Many contemporary accounts consistently misspelled

[14]Slaughter, *From Fortress to Farm*, 67.

Heart River and Butte and used St. Adair for Adair, Rosseau for Rosser, Fort McKeen for McKean, the Gaul for Gall, and numerous variations for Pryor's Creek. Other than Gall, I have changed all of these. Many of the communities mentioned for geographical identification didn't exist in 1872; rather than continually write "present-day" this or that, I simply name the city. Finally, 140 years ago many writers and newspaper editors appeared to loathe paragraphs and periods; I have added both for ease of reading.

* * *

And with that, let us return to the optimistic days—for the army and Northern Pacific Railroad at least—as the spring of 1872 began, well before the company's surveyors had been driven out of the Yellowstone valley.

PART I

To the Yellowstone

May–August 1872

Eastern Survey

Col. David S. Stanley

There is little reason to believe that Ohio-born David Sloane Stanley (1828–1902), commanding the Dakota District in Hancock's Department of Dakota, was an easy person to know. While able, intelligent, and with a strong bent for the sciences (ninth in his West Point Class, 1852), he was also a stickler for detail, heavy-handed, introspective, morose, and likely a social bull in a china shop. Stanley took a perverse pride in not getting along with correspondents, and he suffered bouts of depression and uncontrollable drinking, the latter after periods of intense pressure.

Of medium height and almost frail by 1872, he had few army friends except for those who—like himself—were political outcasts in the army's hierarchy.[1] Stanley's Fort Sully posting was no accident. It was isolated, and Gen. William T. Sherman, who held Stanley to blame for not trapping Hood at Atlanta in 1864, sent Stanley there in 1867, a post (prison for Stanley) without walls near present-day Pierre, South Dakota.[2]

Yet Stanley was also well liked by civilians including Father Pierre de Smet and the legendary riverboat captains Joseph LaBarge and Grant

---

[1]The colonels at upper Missouri posts—Thomas Leonidas Crittenden, Philip Regis de Trobriand, William Babcock Hazen, and Stanley—were among those who had managed to offend Sheridan, Sherman, Grant or—in de Trobriand's case—all three. With few civilian opportunities and protected by seniority, these officers invariably played it safe and watched the years pass.

[2]Stanley's letter, Northern Pacific Railway Company records, "Secretary's Department, Unregistered Letters and Received and Related Records, 1864–1876," microfilm M-459, Minnesota Historic Society, St. Paul.

Marsh. The latter and Stanley got along so well that Marsh named his Missouri steamer, the *Josephine*, after Stanley's oldest daughter. Nowhere are Stanley's two sides better seen than in his posthumous *Memoirs*. In these, the first sections concerning his boyhood can only be described as charming; the reader is taken for a fascinating ride through a northern Ohio more frontier than farmland. However, after the first phase of the Civil War, Stanley's tone is typified by anger, second-guessing, gratuitous comments, and self-justification for his every action.

Sherman is supposed to have remarked that Stanley had "no dash," which is precisely what made him an excellent defensive commander. Stanley, personally taught by George H. Thomas (the "Rock of Chickamauga") in the use of combat field artillery, demonstrated his personal strengths again and again, especially during the '64 Atlanta campaign. However, when he became a corps commander near the campaign's end, Stanley perfectly followed Sherman's orders to destroy railroad track, but with his troops in position to block General Hood's retreating army (think the Falaise Pocket in Normandy), Stanley continued ripping up tracks, never realizing how close he was to fame.

Sherman was disgusted, declined to take Stanley's corps on his March to the Sea, and in the decade that followed he saw to Stanley's remote postings. By 1872 the miserable Stanley was beginning to drink. There is reason to believe that his wife considered leaving him and returning to her family's home in Carlisle, Pennsylvania.

In the spring of 1865, Sheridan was sent to Texas with 25,000 combat veterans to pressure the French-controlled Mexican government and, if necessary, to invade Mexico. Sheridan was given *carte blanche* to select his own officers; in essence, his own "all-star" team. Among others were Stanley and Custer. What saved Stanley's career went back to 1848 when he and Sheridan coincidentally met in Ohio, on their way to West Point. If the two were never close, there *was* this special bond, immeasurably helped by Sheridan's respect for Stanley.

Not hurting Stanley was the birth of his only son in September 1872 when Sheridan was coincidentally touring Dakota posts, including Sully. Stanley had arranged that, in the event of a boy, the child would be named David Sloane Stanley (Jr.). With Stanley's wife too weak to see Sheridan, their daughter Josephine brought the infant for Sheridan to see. Nervous, the twelve-year-old, when asked the child's name by the great man, blurted out, "David *Sheridan* Stanley"—and as such he was baptized. A suspicious mind might believe that Stanley's wife, who herself came from a military family and was aware of her husband's social failings, had more than a little to do with the "accident."

* * *

While the 1871 eastern Yellowstone survey had met all the Northern Pacific's goals, its military commander, Maj. Joseph N. G. Whistler (a first cousin of artist James Abbott McNeill Whistler) had added some five days of extra marching because of his inability to read a compass after midday. Whistler's drinking and leadership (or lack thereof) infuriated the Northern Pacific's lead surveyor, Thomas Rosser, as well as an officer close to Stanley, Capt. Javan Irvine. There is no way that, following the survey's conclusion, Irvine, a delightful gossip, wouldn't have informed Stanley of Whistler's failings. Not that it was necessary. Whistler's drinking became public knowledge after Rosser leaked the story to a Minneapolis newspaper, eliminating Whistler from consideration to lead the '72 survey.

In 1872, Stanley was the logical commander. With a host of other combat-tested individuals from which to choose, Sheridan needed a steady commander who was loyal to him. Other than his 1864 run-in with Sherman, Stanley's fighting qualities were known; his caution was an asset for an army that couldn't afford to assign him cavalry support.[3]

* * *

In late March 1872, Spotted Eagle, just under forty, a Sans Arc, Teton (or Lakota) Sioux chieftain, was sent by Sitting Bull to meet with Stanley.[4] Spotted Eagle told Stanley that the surveyors and their army escorts would be met by force.

---

[3]When Stanley requested cavalry support, he was refused. The reason was the lack of nearby cavalry units, combined with the cost/logistical difficulty of riverboat travel, the only possible way in 1872.

[4]Virtually nothing is known about Spotted Eagle, in part because there were three Teton Sioux Spotted Eagles all of about the same age. This Spotted Eagle led almost two hundred warriors at the Little Big Horn and later fled to Canada with Sitting Bull. He shows up in an 1886 Indian census, but no other information is available. A photo of him survives, but in this less than flattering shot (by the famed L. A. Huffman) he looks confused and intimidated, partially because of a "wandering" right eye.

Because Indian threats sold newspapers, accounts appeared across the county. One of the first and most detailed descriptions is in the April 30, 1872, *Yankton Herald* (South Dakota). Stanley wrote Thomas L. Rosser, who would lead the eastern surveyors, making it clear that Spotted Eagle's threat needed to be taken seriously.

Stanley to Thomas L. Rosser, April 4, 1872

Dear Genl:

I wrote you some weeks ago, but as I heard of you [being] on the Missouri one week ago I judge you have not yet received my letter. My object was to ask you to interest yourself in getting a position for [former Confederate general] Lomax on the NPRR.[5] He is anxious to get employment, and if Lomax is as correct now as he was before the war, I know he would be a useful man for the company. Mr. Jay Cooke promises to help him, but the promises of moneyed men cannot be counted better than "just in promises."

I want to come and see you by some of the first boats, if I can get away.

I had a long talk with Spotted Eagle, a chief and head soldier, Sans Arc Sioux, from the Yellowstone, a few days ago. He was emphatic in his opposition to the Rail Road, and declaims he would [not] listen to new proposition[s] on the subject, [but] that he would destroy the road, and attack any party that tried to build it.

I tried to make him understand the fatal consequences to his people he would bring about by this course. He said he did not care, "it was life of death and he would fight it out."

Tis best you should know. There is no danger to small [surveying] parties east of the [Missouri] river. But there will be constant danger west of it. And it should be impressed on all persons that eternal vigilance is the price for keeping their scalps.

Write and please let me know what [are] the prospects for Lomax.

Source: Northern Pacific Railroad, Secretary's Department: Unregistered Letters Received and Related Records, Undated and 1864–1876, microfilm M-459, Minnesota Historical Society, St. Paul.

---

[5]Lunsford Lindsay Lomax (1835–1913) and Rosser fought together as Confederate cavalry generals in the Shenandoah valley (1864). A West Point graduate (1856), Lomax served with Stanley on the frontier before the war. By not approaching Rosser directly, it raises more questions than answers about his relationship with Rosser, and ultimately he was never hired by the NPRR.

Yankton Herald, April 30

COL. DAVID S. STANLEY MEETS WITH SPOTTED EAGLE THE INDIANS AND THE NORTHERN PACIFIC R.R.

A dispatch from Col. D. S. Stanley, commanding at Fort Sully, Dakota, April 7th, states that several weeks prior to that time, one hundred lodges of Indians under "Spotted Eagle" and "Soldier" of the Sioux [Sans] Arcs, appeared at the agency near [the] Big Cheyenne [River].

The chief has been a big man among his people for about eighteen months, having won his distinction from having led a war party which killed all of a party of twenty-nine Crow warriors last year. He is now suffering from a lung wound received from the Crows. The party came from [the] Rosebud, near its junction with the Yellowstone, their object being to trade and return to their country. Spotted Eagle refused to take rations or any presents while at the agency, but was not disinclined to talk, and was well behaved and orderly in his dealing with the traders and other white men.

He requested Colonel Stanley to hold a council and talk. When with the [Indian] agent and [Army] officers he appeared to be very well informed, and knew all about the threatened invasion of the Black Hills, but did not allude much to it, as he stated he had been told the Great father at Washington would prevent the occupation of that country by white men.[6] He objected to the telegraph line terminating at Fort Sully, and spoke of the Northern Pacific Railroad, and the intention of running it through his country.

Neither himself or the Sioux [Sitting Bull] who authorized him to speak for his people, had consented to this, and they would never listen to any proposition to this effect. He would fight railroad people as long as he lived, and would tear up the track and kill its builders. Upon its being stated that the railroad would be built despite his opposition, and that such a course that he proposed would bring ruin on his people, Spotted Eagle admitted the truth of that, but the driving off of the buffalo was death to his race, and he would fight, knowing he would be beaten.

The agent advises that when the Railroad crosses the Missouri River it must look for such hostiles, and he does not believe that negotiations can do anything with this wild race. At the same council alluded to in

[6]In the winter of 1871–72, a rumor of gold in the Black Hills almost set off a stampede, but the prospectors were, in fact, blocked by the army.

the dispatch, other Indians spoke defiantly, but Spotted Eagle talked very calmly, without passion, and Colonel Stanley believes all he said.

Thomas L. Rosser

Thomas Lafayette Rosser (1836–1910), the man chosen to lead the Northern Pacific's eastern Yellowstone survey team was, at best, a weak civil engineer, but nevertheless the absolutely right man for the job. Standing about 6'2" and weighing 230 pounds, Rosser led by example. After almost five years at West Point, the standard curriculum just before the Civil War, Rosser, who roomed next to George Armstrong Custer for four years, resigned to join the Confederate Army.

After a brief period in the artillery, his charismatic leadership caught the eye of J. E. B. Stuart and by mid-1862 he was the colonel of the Fifth Virginia Cavalry. Rosser claimed to have been in more than one hundred skirmishes and battles and to have killed at least a dozen men in hand to hand combat. By 1865 he was a major general. At Appomattox, Rosser's cavalry was the last unit to break free of the Union encirclement—led by Custer. A man with a strong personal sense of honor, in May 1865 Rosser let himself be captured versus what he felt was the humiliation of surrender.

Following a wartime romance, in the late 1860s Rosser found himself with a growing family and no income. In St. Paul in the winter of 1869–70, Rosser caught the eye of Gen. Ira Spaulding, an ex–Union combat engineer who was leading the Northern Pacific's construction efforts. Despite unhealed war wounds, Spaulding held no grudge and hired Rosser in time for the latter to attend the railroad's February groundbreaking ceremony. Rosser was quickly promoted, and when a leader for the 1871 eastern Yellowstone survey was needed, he was the natural choice. Better yet, he soon discovered he was adept at working with the press, helping his own career as well as giving the railroad positive publicity.

Rosser's 1871 work was considered outstanding because he had found a route through the Dakota Badlands—one remaining mostly in use today. This had been done in the face of physical adversity including a massive prairie fire, Whistler's weak leadership, and Rosser's personal battle with the "Dakota Quickstep" (that is, piles). In 1872 no one else was considered to lead the surveyors.

Throughout the 1872 survey Rosser kept a field diary, penned letters, and in the 1890s wrote an article, "The Death of Adair," none of which have been published.[7] In the article he describes the surveyors' difficult lives and the fighting, including using Gatling guns against mounted Plains Indians. Yet perhaps the most interesting of all Rosser's writings is an October 11 "report" in which he, clearly still traumatized, tells how he barely survived an ambush led by the war chief Gall.

Rosser wrote the unpublished "Death of Adair" in the mid-1890s. Regarding the 1872 surveys, it is perhaps the single most important post-survey document to be found. Despite melodramatic Victorian passages and personal prejudices, Rosser's manuscript relates numerous aspects of the survey not found elsewhere and also confirms stories otherwise subject to question. Equally important, it must be remembered that Rosser was a combat-hardened officer, and his accounts of Indian fighting are outstanding.

Rosser gave the article Adair's name for the simple reason that Lewis Dent Adair was the first cousin of Julia Dent Grant, wife of the president.

This section, the first of three parts of Rosser's article, is easily his stylistic worst, his nineteenth-century ruffles and flourishes on full head-shaking display. (Was this the reason he never found a publisher?) And while his account of crossing the Sheyenne River technically took place in 1871, few sections more ably illustrate the physical problems that were encountered.

The Death of Adair (excerpt)

The "Tide of Empire" in its western march has swept over the wild hunting grounds of the "noble red man," destroying his flocks and herds of antelope, deer, elk and buffalo; breaking his proud spirit by hunger, and bringing him in humiliation and sorrow to the feet of the despised white man, who prescribes his daily ration and sets limits and bounds to his fishing brooks and hunting grounds.

One, in riding over the plains of the Upper Missouri, frequently passes over the traces of once large "Tepees Cities," many of them miles in area evidencing a population of tens of thousands. These marks are circles

[7] It is difficult to believe that Rosser did not write some sort of final report for the 1872 survey, but I have been unable to locate it.

of stones, each weighing about 20 pounds, placed around to hold the bottom of the hides which form the covering for the tepees poles. These stones are, by time, half buried in the ground, showing that they did their service many years ago.

Where are the souls who once peopled these plains? The bleached bones of the buffalo are still to be seen in great quantities; even the large piles of crushed bones, just as they were dumped from the soup pot, are numerous about these "tepees cities." This once active population of warriors and hunters has disappeared, and these plains are lifeless and still. The beaten trails of the buffalo, running like dark threads through the green prairie, and worn by the traveling herds while passing in single file to and from the water pools, are fast growing up and the nutritious game grass comes and goes with the seasons, untrampled and uneaten.

I first crossed the Dakota prairies in the spring of 1871. I was sent out by the Northern Pacific Railroad Company to ascertain if the route, directly west, from Georgetown [Minnesota] on the Red River of the North, to the Missouri, was practicable.[8] This country was uninhabited, and with the exception of an occasional visit from Indian hunting parties, its quiet and stillness was never broken, except by the wail of the coyote.

Mr. George A. Brackett of Minneapolis procured for me the services of Pierre Bottineau, and with one assistant, Major McGregor, and two common two horse wagons, we set out from St. Cloud, Minn. on our expedition, April 21st.[9] The major, myself, and guide were on horseback, one wagon was loaded with forage exclusively, the other carried rations and baggage as well as forage. It was too early for grass, and as I calculated, to move rapidly it was necessary to feed the animals well.

I crossed the Red River at Fort Abercrombie, and marched via the Fort Totten staked trail, over the level prairie until I reached the latitude of

[8]The current Georgetown, reflecting a change in the Red River's course, is about a mile east of the Red and a dozen miles north of Moorhead, Minnesota.

[9]George A. Brackett (1836–1931), a native of Maine, was one of the first mayors of Minneapolis. During and after the Civil War, he served as "provisioner" for the heralded First Minnesota, other wartime regiments, and army posts in Minnesota and Dakota. By 1869 he was supplying and equipping all Northern Pacific survey teams. He and Rosser became good friends, later engaging in side businesses together. The legendary Pierre Bottineau (ca. 1814–95) was possibly the leading scout in the Dakotas during this period. A town and county in North Dakota were named for him later.

Georgetown, then turned off due west. The melting snow had flooded all the streams and there were neither fords nor bridges and when I reached the Sheyenne River, a stream 200 feet in width and rapid current, I vainly wished for a pontoon train, [or] a few boards to make a skiff, a "dug out" or a few logs for a raft. But there was nothing; a few small willow switches growing along the bank of the stream were the only semblance of wood in sight.

Pierre Bottineau, my guide, was a half breed Sioux. He was an intelligent fellow and had lived a long time on the plains. As soon as we reached the bank of the river he began unloading the baggage wagon without saying a word as to his intentions (he may have supposed that I divined them) but as soon as the wagon body was empty, he asked for assistance to remove it from its bed and sit it on the ground. He then took the fly of my tent, spread it on the ground, and then had the wagon body lifted and set on the tent fly. He then drew the fly up, and secured it with cords and lines tightly around the box. He then had the wagon box, thus prepared, lifted to the river, then took a tent pole and one end of his rope and entered his improvised boat. He pulled upstream by the small willows until he reached the desired point in the bend of the river, then he pushed off into the stream.

[B]y a vigorous use of the tent pole as a paddle, he drove his boat to the opposite shore having, of course, drifted with the current several hundred feet, but he then pulled up along the shore until he reached the desired point. He then secured his end of the rope to the young willows, and the other end held by one of the teamsters he pulled himself and [the] boat back to our side of the river. We then had a 'rope ferry' which soon landed us forage and baggage safely and dry on the west shore of the river. The horses were then made to swim across and after this detention of not more than two hours we were able to resume our march, confident in our resources for any emergency.

Following the most favorable depressions and drains we reached the Missouri River, via the valley of Apple Creek, at the point where the enterprising and flourishing city of Bismarck now stands. My report being favorable this route was adopted, but we were still uninformed as to the character of the country lying between the Missouri and Yellowstone rivers. And although a good line could be secured to the Missouri, it would not be safe to build without being sure of the practicability of the route beyond. Therefore a reconnaissance of this country was ordered in the fall of '71.

Under orders from General Hancock an escort was ordered from Fort Rice on the Missouri River to accompany the survey, and with a full surveying party with forage and rations for sixty days, and an escort of about six hundred infantry, a section of artillery, and ninety mounted Indian scouts, all under the command of Gen. Whistler, an able and experienced officer,[10] we marched out of Fort Rice on the 9th of September. [We] proceeded up the Missouri to the mouth of Heart River and began the survey, which we pushed rigorously to the mouth of Glendive Creek on the Yellowstone. This region was at that time filled with game—antelope by the thousands were seen daily, quantities of deer and elk supplied us with fresh meat, and a few [old male] buffalo gave us an opportunity of trying the speed of our saddle horses and our skill with the revolver at close range. But in rapid motion we returned to Fort Rice about the middle of October, the exploration having been regarded as quite a success.[11]

Rosser Diary, June 28–August 15

JUNE 28. [Fargo] Party of young men from hunt occupied my tent. Met Mr. Morison on business of the company. Arranging the general routine business of office, etc. D.C.L. [Daniel C. Linsley] and [board of directors] party en route for this place.[12]

---

[10]Rosser must have assumed that this account would be published because, in fact, in 1871 he clashed frequently with the often inebriated Maj. Joseph Nelson Garland Whistler (1822–99). Rosser returned from the survey so angry that, using a pseudonym, he wrote to a Minneapolis newspaper about Whistler's drinking, and the letter was carried by the wire services to newspapers throughout the country.

[11]Rosser is understating the results of his survey, a thorough technical triumph with only a handful of Sioux spotted. To a considerable degree, Northern Pacific overconfidence regarding the Sioux was caused by the success of both Rosser's and the western survey.

[12]Daniel C. Linsley (1827–89) was the Northern Pacific's second ranking engineer. His Vermont family was close to the Smiths. He likely joined the railroad with the understanding that he would replace the aging Edward C. Johnson. However, with Johnson's 1870 retirement, Linsley was passed over in favor of Milnor Roberts, Cooke's first choice. In charge of construction and engineering in Minnesota in 1872, Linsley was technically able but heartily disliked by all the surveyors.

JUNE 29. At Fargo arranging to take [Board of] Directors to Missouri River. Very warm and in afternoon rained hard. No wind. [Our] Track layers struck for more wages.

JUNE 30. Very warm. Party of young gentlemen with us and fearfully in the way. Had an interview over telegraph line with D.C.L. Arranged program for expedition over St. Paul & Pacific RR. Went to church. Spent a day pleasantly with dear Lizzy.[13]

MONDAY, JULY 1. Spent day very pleasantly at home with my sweet family. Made preparations to send D.C.L. and Governor Smith to St. Cloud and to take [Chief Engineer] Roberts to Missouri River.[14]

JULY 2. At Fargo making arrangements to receive D.C.L. and Gov. Smith. Finishing bridge [across Red River]. At work on house. Boring. Will lay track and ball aligning a little.

JULY 3. At Fargo engaged as on yesterday.

JULY 4. At Fargo making arrangements for Governor Smith and party. Arrived 6 P.M.

JULY 5. Gov. Smith and party. Rode in engines to the front. Made out estimates and discussed various points of interest on work classification determined.

JULY 6. In company with D.C.L. and Mr. Roberts. Rode to end of the track by rail, thence by freight wagon to point 2 miles west of big coulee where we changed horses. Took dinner with Clark[15] and drove to Lake Eckelson where spent night.

JULY 7. With fresh team we drove [west] to James River. Stopped a few minutes then drove to 10 Miles Spring. Took a fresh team and drove to M[——]'s camp near Long Lake. The drive a long and tiresome one.

MONDAY, JULY 8. Drove from Long Lake to Stevens' camp where we took dinner. Thence we drove to Missouri River reaching there about 2 P.M.

---

[13]Betty Barbara Winston Rosser (1843–1915), Rosser's wife, also known as Lizzy.

[14]Governor Smith is J. Gregory Smith, the Northern Pacific's president and former governor of Vermont. W. Milnor Roberts was the NP's chief engineer and Linsley's superior. Unlike his relationship with Linsley, Rosser got along well with Roberts.

[15]G. A. Clark was a surveyor.

Went aboard the [Northern Pacific owned] steamboat and ran up the river beyond the mouth of the Heart River. Thence down river to Apple Creek and lay off Eckelson's camp for the night. Mosquitoes fearful.

JULY 9. Rode over the country along Buck Creek and with Mr. Roberts rode over to Heart River. Saw a great many antelope. Much pleased to have Mr. Roberts see the valley. He appeared to be much pleased. Spent afternoon on boat and night on prairie east of [the] Missouri.

JULY 10. Spent fearful night on prairie. Put the horses in an old house, kept smudges [pots to thwart mosquitoes] going all night. It rained and the mosquitoes were fearful. Left Missouri River about 6:30 A.M. and drove to Long Lake for camp.

JULY 11. Drove to the Dakota Spring where we examined the country with a view of changing the line. Thence to Olmstead's camp, changing horses at Patterson's camps we drove to James River for dinner thence to Lake Eckelson where we spent the night.

JULY 12. Left Lake Eckelson 6 A.M. and drove to Sheyenne where we changed horses, thence we drove to [——] where we changed again. Thence to the end of the track about 3 [or 8] miles east of Goose Creek. Reached Fargo about 5 P.M.

JULY 13. Mr. Linsley and Roberts went east; I remained to arrange the party for the trip. Fixing up matters generally. Writing letters, paying my bills, etc.

JULY 14. Got the party ready to start and loaded them on the train. Horses, wagons, etc. also put aboard and sent to the end of the track that they may get an early start tomorrow A.M.

MONDAY, JULY 15. Party left this A.M. for the Yellowstone. Arrangements all right and party got good start. Making arrangements to follow.

JULY 16–17. In Fargo fixing up matters for the trip to the Yellowstone.

JULY 18. Turned over work to Col. Gaw.[16] Gave him orders, etc. and made all arrangements. Fargo tomorrow. *Sad, sad indeed.*

---

[16]William B. Gaw (1821–90) was an engineer who was primarily assigned to construction, not surveying. He is often referred to as Colonel based on his rank while leading the Sixteenth U.S. Colored Infantry.

JULY 19. Left Fargo 6:30 A.M. Took train at end of the track and drove to Clarke's on summit for dinner. Gave him instructions about the work. Had some words with Gaylord about pipe, etc. Drove to James River where we spent the night. Dreadfully blue.

JULY 20. Drove to Rose Lake. Grass on prairie very much burnt since ten days ago.

JULY 21. Reached the Missouri River 11 A.M. and arranged to have horses shod, wagons fixed up, etc. Took dinner and had boat fired up and ran down to Fort Rice. Mosquitoes fearful beyond description.

MONDAY, JULY 22. Spent greater part of day at Rice. Purchased several small articles from Q.M. and got 24 carbines from] Col. Crittenden. Ran up river leaving Rice late in P.M. Dinner with Genl. Crittenden.[17]

JULY 23. Ran all right [on paddle wheeler] last night. Reached Apple Creek [Bismarck] about daylight. Mosquitoes perfectly fearful. Got up early and walked over to Eckelson's camp. Got horses for the [——]. Spent day in looking over the ground and setting the site for the post. D.C.L. returned to Rice on boat.

JULY 24. Finished transferring freight from river to new storehouse and put all wagons and harnesses in a state of repair. All things being ready I am anxiously waiting orders for the march which I expect noon tomorrow. Have been writing to my darling wife.

JULY 25 Orders came about daylight. The boat had already arrived and was waiting to ferry my teams, wagons, etc., over. We were not quite ready but went to work well before breakfast and by ten A.M. loaded

[17]Thomas Leonidas Crittenden (1819–93) was a colonel in 1872 and a Union major general during the war. He was the son of Senator John J. Crittenden of Kentucky (an 1860 presidential candidate), and his brother, George, was a Confederate general. After being dismissed by Grant during the war (partially for co-leading the retreat at Chickamauga), he was given a colonel's rank by Andrew Johnson. Crittenden and Rosser became close, the latter unsuccessfully attempting to help when Crittenden's son (John J.) was forced to resign from West Point. Despite losing an eye, the son stayed in the army and was with Custer at the Little Big Horn. His body was never "officially" found, but his bloody underwear was discovered in the Sioux camp where he was undoubtedly tortured to death. The son's death broke Crittenden emotionally, and he received eastern sinecures for the remainder of his career.

and by 2 P.M. landed on the west shore and went into camp. One man badly hurt whilst unloading.

JULY 26. Left camp 6 A.M. and marched directly at Buck Creek. D.C.L. very troublesome about selecting roads and looking after latter details greatly to the annoyance of us all. Reached camp 4 P.M.; wood not very convenient, water good [from] spring. I write [*sic*] a note to my darling wife.

JULY 27. Did not leave camp until about 8 A.M. Sent out to look for General Stanley and before learning where he was, we started out, had a hard drive, very hot and bad traveling. Halted to wait for Stanley and I went out and killed an antelope. Horse ran away and I had to walk back to camp.

JULY 28. Went into camp last night about 4 P.M. Began raining just as we began putting up tents, ours was blown down. Did not rain a great deal. Slept sweetly. Had pleasant dreams of Lizzy and the children. Left camp this A.M. 5 o'clock. Marched only 4 miles through a good country.

JULY 29. Made short march yesterday of only four miles. Left camp this A.M. and marched via big bend of Heart River and encamped on Whistler Trail opposite mouth of Mud Creek. Thunder shower threatened but no rain. Messenger went back to Missouri River and I sent letter and telegram to my precious wife. There are times when she doubts my love. This is wrong, for I love her devotedly.

JULY 30. Left camp 5 A.M. The weather cold and the marching good. Reached Heart River 9 A.M. Had some trouble getting over but finally crossed and moved off about 1:30 P.M. Marched to Spring Creek and encamped for the night. The stream looking beautifully. Several animals killed by the party. One elk killed. Am very sad. God bless Lizzy.

JULY 31. Did not leave camp until 9 o'clock. Several wagons had broken and rations had to be issued. We managed about 12 miles and encamped on Ghost Creek. Went up on Heart Butte. Very windy below but calm on top. Killed four rattlesnakes. Got specimens of petrifications. Mail got in, nothing for me.

AUGUST 1. D.C.L. and myself took the scouts and rode over to Heart River and up that stream to Young Men's Creek. Saw a large herd of elk but killed none. Crossed creek with one accident and camped on head waters thereof. Feeling badly and unhappy. Don't know why, but fear that a great domestic trouble awaits me.

AUGUST 2. Left camp 5:30 A.M. and made a hard march to Sully's [1864] crossing of Heart River. Crossed without difficulty and began running a line south 79 [degrees] west. Country rough and progress slow. Mind still filled with strange fancies. I love my wife and hope she will appreciate me.

AUGUST 3. Continued running the line of one of the branches of Heart River. Watered, cut grass and took dinner about 12 noon. Then ran off of Heart River crossed a ridge and encamped on another fork of same river. Wood and water in abundance. Spent a very disagreeable night.

AUGUST 4. Day very warm and disagreeable. Saw some bad lands but got through them well. About 12 noon came on the South Fork of Heart River. Bad crossing and it was decided to encamp on it for the night. Got over without trouble and spent a very quiet Sabbath afternoon. Not feeling well.

MONDAY, AUGUST 5. Broke camp as usual and I rode ahead to examine the country and to see where we could strike the Bad Lands. Found that we were too far north and turned off south. Went into camp and I with Indian scouts rode to Little Missouri to look up the road. Spent the night on Little Missouri in bivouac.

AUGUST 6. Left bivouac at daylight this A.M. with Mr. Eckelson and Mr. McDonald rode up the Little Missouri River and got on a high butte. Overlooked the country south, with the view of judging of its practicability for a rail way route. Thence returned through the Mauvaises Terres [Badlands] to camp and moved five miles to new camp.

AUGUST 7. A strange dream last night. Left camp very early and ran off south to take the line which I explored yesterday. Found the line practicable and a good wagon road. Made a long march considering the country and encamped on the Little Missouri River. Storm came up just as we were getting into camp but fortunately we got under our canvas before it came off.

AUGUST 8. Rained violently last night and it has continued all day. D.C.L. and myself went out to look up route for wagons early A.M. Scout sent out did not return when expected and party was sent for him. I rode out to look up wagon road and when returning my horse down a hill he fell and hurt my arm and shoulder quite much. Cut himself on shoulder.

AUGUST 9. Cold and wet last night. Had fire in my tent. Left camp as usual and found the road fair. Reached the Little Missouri about noon and found it too high to ford and we were compelled to [remain] on East bank and wait for it to recede. Found large quantity of fossils. Large fossil [tree] forms of several hundred feet long as they lie exposed in the hills.

AUGUST 10. Spent the day in camp. Not able to cross the Little Missouri River. Had a map made of line as far as run and cables prepared to assist in crossing the river tomorrow. Looked at fossils and slept some. In the midst of the fossil [——] specimen 5 to 8 feet in diameter and 300 feet long.

AUGUST 11. Descended a precipitous hill to the river and finding it had fallen so as to make the crossing appear easy I desired the ropes dispensed with. But D.C.L. insisted and I left, in disgust with the line party. Found a tolerable line cut out of the [Little] Missouri valley. Country fair with wood and water. Encamped near large spring.

MONDAY, AUGUST 12. Buried poor [teamster] Pettit this A.M. He died this morning about One o'clock of consumption. We had a coffin made of a wagon box and gave him quite a decent burial. General Wallen reading the services. Reached the waters flowing to the Yellowstone. Camped on high hill overlooking same.[18]

AUGUST 13. Spent all the A.M. in getting down the bluffs of the Bad Lands into O'Fallon's Creek. Water scarce and bad. Camped on O'Fallon's Creek. I took the scouts and explored west and found good country and water. Made a hard ride and did not return to camp till 8:30 P.M. Tired and sleepy.

AUGUST 14. The direction of the survey was turned south to get us into the country I examined last night. And after getting out of the deep valley and Mauvaises Terres in which we were encamped, we found a

[18]Maj. Henry D. Wallen (1819–86) was a brevetted brigadier general. His being on the survey was "punishment duty" ordered by President Grant for an incident dating back to the late 1840s. See *Jay Cooke's Gamble,* 119.

good country. Reached my spring about 10 A.M. and at 2:30 P.M. found water in great abundance and encamped.

AUGUST 15. Took the Indian scouts and rode over the country in advance of the escort. Followed down the creek on which we were camped last night to its mouth where it emptied into a long stream of flowing water with broad level valley covered with heavy cottonwood timber. Hence rode over the Bad Lands and returned.

Journal of the N.P.R.R. Surveying Expedition from Missouri River to Powder River, July 26–August 17

FRI. JULY 26TH Struck camp on west side of Missouri River at 6 A.M. and took up the line of march west—accompanied by Capt Moore—twelve scouts, and a small detachment of soldiers as escort—Marched twenty miles and camped on Buck Creek near a fine spring the water of which had a slight taste of sulfur—Heavy shower with strong wind in the evening three of the tents were upset—

SAT. JULY 27TH Clear & very warm—Struck camp at seven and marched two miles where we halted to await for our escort from Fort Rice under command of Gen. Stanley. Escort arrived at one o'clock. Marched between two and three miles in the afternoon. Went into camp about four o'clock.

SUN. JULY 28TH Rained in the morning until eight o'clock. Struck camp at six and marched two miles to the North fork of the Cannon Ball where we went into camp for the day—Very warm after the shower.

MON. JULY 29TH Clear & warm—Struck camp at five o'clock & marched about fifteen miles through quite a heavy rolling country. Went into camp at 1½ on the border of a creek where we found good water although rather warm—& plenty of good grass for the cattle.

TUES. JULY 30TH Clear & chilly in the morning with a *raw* wind—Struck camp at quarter to five—Reached Heart River at 10 o'clock which we forded without accident—Got all the teams across at one o'clock—Marched to Spring Creek and went into camp at 4½—Water very good—supply of grass fair—Distance marched about eighteen miles.

WED. JULY 31ST Clear & chilly with strong westerly wind—Rations had to July 31st be served to troops which delayed us so that we did not march until 8 o'clock—Roads very good and country much more level than we have been previously traveling through—Made good time on the march—Passed Heart Butte between ten and eleven—Went into camp at 2 o'clock on the bank of a small creek—Water good—grass abundant—Marched 13 miles.

THURS. AUG. 1ST Clear & chilly in the morning—warm after nine o'clock—Broke camp at 5 o'clock made fast time on the march until ten o'clock when we struck a branch of Heart River & were delayed some time—Went into camp about six miles from creek—Had to leave the trail and go about 1/4 of a mile to the north for water—grass plenty—no wood in the immediate vicinity—good supply about a mile and a half west—Marched about 15 miles

FRI. AUG. 2ND Clear—Air very sharp early in the morning but came out very warm in the middle of the day—Crossed Heart River between 11 & 12 o'clock—Commenced running line after lunch—ran three miles of chronometer line[19]—Went into camp at 4o'c—Found a splendid spring near camp—Supply of grass fair—Marched about 19 miles

SAT. AUG. 3RD Clear & warm—Struck camp at 5 A.M. & marched until 2½ making nearly 22 miles—Marched through a level country Camped on Sandy Creek—Wood plenty also water & grass—Chronometer party ran 17 miles.

SUN. AUG. 4TH Clear & warm—Broke camp at 5½ A.M.—Crossed the south branch of Heart River at noon and went into camp on the north side. Chronometer party ran eleven miles—Indians reported. Gen. Rosser sighted two ahead—Train marched 15 miles—

MON. AUG. 5TH Clear & warm—Struck camp at five o'clock and marched until twelve when we were obliged to halt as the chronometer party report impossible country ahead—Gen Rosser, Eastman[20]—MacDonald

[19]Chronometers were used to measure time and location.

[20]Frank Eastman (d. 1884) was the son of the general and famed artist Seth Eastman, whose paintings are on the walls of the Capitol's Rotunda. Eastman, a surveyor since at least 1870, had his father's artistic skills, which were utilized to sketch landscapes.

& Eckelson[21] accompanied by the Scouts rode forward to the Little Missouri to find a road. We marched fifteen miles to-day.

TUES. AUG. 6TH Clear—Laid in camp until 2 o'clock waiting for return of Gen Rosser & party—On their return we struck camp and marched south two miles and camped again—The reason for moving was to get better water than we had at our previous camp, which was very poor.

WED. AUG. 7TH Clear & warm—Struck camp at 5 o'clock and marched south for several miles and then struck South westerly—Had a very rough hilly road We got through without any accident to our own teams, but three of the government teams broke down—Marched twenty miles—Camped on a branch of the Little Missouri, plenty of wood and Water. No grass of any account.

THUR. AUG. 8TH Rainy—Bugle sounded at 2 o'clock did not break camp until eleven o'clock as we had to wait for the scouts who had been sent out to find a trail—Roads very heavy from the rain, and very hilly—The country so rough that chronometer party had to dispense with their wagon and pack the instrument—Marched 7 miles.

FRI. AUG. 9TH Clear & cool. Struck camp at 5—marched through a country less broken than that we have been travelling through for the past few days although still very hilly—Saw numerous coal veins along the road. Reached the bluffs of the Little Missouri at 12 o'clock. The river was too high to cross so we went into camp on the best camping ground that we have had yet. The scenery from our camp ground is very grand—Found some very fine specimens of petrifaction, the finest specimen is an immense fern tree, said by the geologists of the party to be of a species extinct before the creation of man. It was found at the foot of a high bluff that has been worn away by the elements—Saw several Elk—marched ten miles.

SAT. AUG. 10TH Clear—Laid in camp all day as the river is still too high to cross the teams—all of the party passed the morning in collecting

[21] Albert O. "Eck" Eckelson (ca. 1845–1902) was an able, self-taught surveyor and married with three children. A Civil War enlistee, he was badly wounded at Stone River/Murfreesboro and contracted tuberculosis while hospitalized. He was well liked: Custer thought highly of him, too. A lake and a small North Dakota community bear his name.

fossils—Found several very good specimens but the rock in which we found them was very soft sand stone so—that it was not possible to preserve many of the specimens—In regard to the ferns of which I wrote yesterday there is a great deal of uncertainty about them—they prove to be sand stone and if they are not fossil trees are certainly very remarkable formations—Gen Rosser and Mr. Linsley both think that they are ferns but I doubt it as I cannot see anything except the shape that looks like tree.

SUN. AUG. 11TH Clear—Crossed the river without any difficulty and were all across by ten o'clock—Gen Stanley saw three buffaloes probably old bulls—Marched twelve miles camped at Sta. 4599 on chronometer line.

MON. AUG. 12TH Clear—Pettit the clerk for Clark & Waits outfit[22] died last night and was buried this morning—Struck camp at 7½ Went out with chronometer party—Struck the Bad Lands at 2½ o'clock and had to abandon the wagon & pack the instrument—Ran four miles through very rough country—Ran until after six o'clock as we were not in sight of the train and expected every minute to see a messenger to let us know the way to camp—Stopped the line at 6¼ making 16½ miles of a run—Did not find camp until 7½ o'clock; walked about six miles through as rough country as I ever saw and reached camp pretty well used up—Had a very poor camp as there was very little grass and poor water and that was two miles off.

TUE. AUG. 13TH Clear—Did not break camp until past 7 as the scouts had to go out to hunt a road—Staid in bed until 5½—Had considerable digging to get a road down from the bluffs to the low ground—Mr. Linsley had all the engineer corps out picking and shoveling—Did not get to work on the line until noon. Ran until three o'clock when the train went into camp. Have the poorest camp that we have had yet—in the midst of a sage prairie, *very* poor water and no grass—Ran 5 miles—Scouts sighted 25 Indians.

WED. AUG. 14TH Clear and warm—Broke camp at 5 o'clock—Ran the line south west through the hills for two miles and then ran westerly through a rolling prairie. No water for the horses from the morning until

[22]The Clark & Waits company managed the civilian teamsters.

four o'clock in the afternoon when we camped on the edge of a creek on a sage brush prairie—Marched 15 miles.

THUR. AUG. 15TH Clear & warm—Were delayed two hours in the forenoon with the line waiting for the train to catch up—Came on to O'Fallon's creek about two o'clock and ran along with it north for six miles. Crossed creek five times—valley timbered with large cottonwoods—Ran 14½ miles.

FRI. AUG. 16TH Cloudy & cool—Pickets attacked by Indians at daybreak casualty list, one mule wounded—Followed the course of O'Fallon's Creek all day—Were delayed again by train being behind—Scouts saw twenty Indians come into our camp after we left it this morning—Ran about 14 miles—Camped on O'Fallon's Creek

SAT. AUG. 17 Clear & warm—Entered the valley of the Yellowstone about ten o'clock—Just as we left O'Fallon's creek the alarm of Indians was given and the train was corralled & a skirmish line thrown out but the Indians did not show themselves—Ran the chronometer line up the valley until four o'clock when we camped—Found plenty of grass for the animals—In the evening Gen Stanley fired several shells from the Napoleon & also from the Gatling guns as a salute.

Source: Manuscript 137, K.13 (B), Minnesota Historic Society, St. Paul.

Letters of Montgomery Meigs, January–July

It is safe to assume that Montgomery Meigs (1847–1931) was hired because his father was the army's quartermaster general. Like so many in his family, "Monty" was a bright, highly able individual whose education included Harvard and graduate study in Stuttgart, Germany. Meigs, who joined the Northern Pacific in 1870, was strong willed, industrious, and frequently took on difficult, dangerous assignments in order to show his independence and self-worth. During the winter of 1871–72 Meigs was with the small forward party of surveyors who lived in tents near where Pipestem Creek enters the James River, a location that afforded protected from the wind, but not the bitter cold.

Meigs was not chosen for the 1872 Yellowstone survey, although he was appointed acting chief engineer in the Bismarck area. Had 1872's construction come close to meeting its goals, his position might have been meaningful. Nevertheless Meigs's letters offer unique and candid insights into Northern Pacific operations, personalities, and life on the very edge of the frontier.

James River, January 14, 1872

We have had lovely weather for the past two weeks. The thermometer has rarely been below freezing, and bright still days that made it a real pleasure to be out of doors have succeeded each other to an extent that has astonished us all. Tonight however it blows great gusts and the tent slapping against the camp table at which I write makes it difficult to see the letters.

The Valley where we are camped is a pretty one. High bluffs on each side enclose meadows about one mile wide from bluff to bluff with the winding river bordered with large trees make a landscape worthy of the artist's pencil. Now great banks of snow hang on to the side hills and fill up the ravines while the untrodden snow which the meadows wear helps to give a real arctic look to the scene. Still the tents nestle under the hill and among the trees. Curling smoke from the chimneys hints at warmth and comfort in them. A branch of the James called the Pipe Stem joins its waters with the James just below our camp and the [planned] R.R. line crosses both streams. We are quite a little colony at this place. The soldiers have a comfortable board house. There is an ice house in process of construction also, as well as a large stable, and three of my tents make quite a town.

We had an accident a day or two ago in one of our tents through sparks falling on the roof, and we did not succeed in putting the fire out till it had made an opening large enough to crawl through. Reid, my transit man is a good fellow in every way and I don't often have to scold, but I find it necessary to get mad [all emphasis in original] with the lazy men sometimes though not often. I don't know whether the men growl at me or not. I suppose they do however. I have gotten a little snow blind lately and find the candle hurting my eyes so I will close.

James River, March 1, 1872

Some of my men have left and so I find myself quite shorthanded. It is a great change from the luxury of [home] to camp life, and I found myself quite repelled when I entered my tent and saw how dirty it looked and forlorn. I immediately began a crusade and had a cleaning day which has much improved the appearance of things. A new man in my party, a topographer named Dohua is a Pole.[23] He was educated in Berlin and in the Prussian army. He remained a year or two, but then [came] into some difficulties with a brother officer, shot him in a duel and had to fly the country, coming to America. He is finely educated, speaks many languages, plays the piano and guitar and sings German and Polish songs with exquisite taste to the accompaniment of the guitar. He is, moreover, an exquisite draftsman and sketches, engraves, and ornaments beautifully. In the evening he sings to us and it is a great treat to listen to the old Polish national songs which he sings. He is a miserable man mentally, however, an old bachelor of forty perhaps.

James River, March 17, 1872

I returned yesterday from a trip forty miles west of here, the Coteau, a high desolate country without a stick of timber large enough to make a Lucifer match, for miles and miles. At this season, swept by the storms that howl over the country and buried in snow it is bleak enough, It took four days to go and come and we slept one night five of us in a little tent nine feet square, the thermometer fourteen degrees below zero. I froze both feet but not badly and on taking off my stockings found them perfectly white. Only the skin was frozen however, and I thawed them out with snow and feel no ill effects. I hope to move out from here in a day or two but cannot for want of wagons or sleigh. It is storming furiously out but our tent is comfortable. I wish you could see how warm and pleasant we manage to make it.

James River, March 31

A total change of program. Eckelson a few days since arrived bringing letters from Gen. Rosser. I was informed that "land grabbers" were enroute for the Missouri to make claims and take possession of the RR crossing before the engineers could get there. I was ordered to turn my party over to him that he might hurry out and anticipate this move

[23] Also spelled "Donah" by Rosser. See my *Union College's Class of 1868*, 167.

and locate a place for the [Bismarck] RR bridge. He left next morning taking all but three of my men and leaving me to enjoy a season of comparative leisure. The next morning the land grabbers appeared and halted for lunch near our camp. They had two wagons and five men and were extremely mysterious about their moves, but finding that Eckelson started ten hours before them they hurried on and the two parties are now racing towards the River. I hope he will get there first and save the company a lot of expense. Gen. R and Mr. Linsley are expected tomorrow and are going through to the Missouri. I may go with them as I am promised the Western Div. of the road.

Missouri River Crossing [Bismarck], April [no date in original]

I have already entered on the duties of my exalted position of Resident Engineer (salary I hope of $ 200 a month). I came out from James River with Mr. Linsley and Gen. Rosser and had a pleasant trip though somewhat under restraint as regards Mr. L our Asst. Chief and virtual Chief Engineer on this end of the road. Mr. L is a peculiar man. He is as kind as a razor and as practical as—I can find no simile appropriate. Every word to [him] must be expressed in language which leaves no room for a possible double meaning. His thoughts and habits of reasoning are in the manner of Euclid, and woe unto the young man, so unfortunate as to make a loose statement. No matter how plain his meaning, looseness of expression must be hunted down and exposed, demonstrated, and shown up, after that conversation may proceed. This is to say the least <u>disagreeable</u>. At times he is quite genial and a moment after as cold and hard as a glittering icicle. But I grow poetical. Having no instruments here but a transit I shall have to confine myself to making a map of the river bottoms and river. When my party is sent I am to locate and construct the R.R. line east. It is a great responsibility for one of my experience, but I shall put on that brazen panoply of proof called "cheek" and push on to victory.

Missouri River Crossing [Bismarck], April 30

I have been hard at work on a survey of the localities at the mouth of Apple Creek where our new town of is to be. It is a pretty place, plenty of timber, a wide flat bottom backed by ranges of hills and bordered by the yellow and crooked Missouri. A good deal of my work was on the other side of the river which is nearly three quarters of a mile wide at high water, but at present not over 1,500 feet across. We have a log canoe which holds six people by squeezing them in rather tight, and to convey my party across required sometimes two and sometimes three trips. We always got wet crossing and had to wade across sloughs and mud flats sometimes up to the waist in water with great cakes of ice two feet thick floating about.

The ice has nearly disappeared now, but came very near causing us a serious accident about a week ago. The river had fallen about a foot one morning when we crossed and all the heavy ice was aground on the sandbars leaving the river quite open. So we got across without difficulty and went to work. About four o'clock I returned to where we had left the boat and found to my consternation that the waters had risen again and was filled with ice which came in packs down the stream at the rate of five or six miles an hour. Here and there huge cakes were aground on snags or sandbars and against such impediments the floating ice rushed and ground with a continuous roar and a resistless force that was not in the least reassuring. To stay where we were without provisions would necessitate our making a trip to Ft. Rice thirty miles below, on foot and with empty stomachs unless the river should fall within morning and that was very unlikely. To spend the night at this time of year in the open air without blankets or shelter was also a severe alternative so I concluded to risk the crossing.

There were six of us in the canoe which we previously hauled a long way up stream to give us a chance to float with the current and still strike our landing place on the opposite shore. We pushed out into the stream, and got along very well till 100 yards of the shore where the floating ice was very thick. We moved fast to escape being crushed against a grounded cake and run into a floating one. Once we barely missed one of these formidable obstacles to navigation which rushed past our stern within a few inches and appeared to have the velocity of a Hudson River steamboat. I could not help laughing at a Swede who was steering the boat and whose eyes stuck out so as we shot past that you could have knocked them off with a stick. A moment after we fortunately got into an eddy which helped us a getting into shoal water and soon after we jumped overboard and pulled the boat to safety, feeling thankful for our narrow escape.

You should see how the people are flocking in here and the new houses are beginning to dot the prairie. There are now perhaps 200 people within five miles of here when six weeks ago there were only three woodchoppers living on the island. There is no one living across the river however and we saw lots of tracks of game. Elk, deer and some enormous grizzly or cinnamon bear tracks quite fresh. Our provisions have run out and we are now living on pork, bread and molasses twenty-one times a week. One of the men shot a beaver a day or two since and we had him for two or three meals, but our cook did not know how to make beaver tail soup [a note in the original says to prepare this delicacy use the meat just above the tail or paddle]. Most of us could not swallow the meat which had a strong fishy or musty taste. A good many of the men however devoured their portions and Oliver Twist like held out their plates for more. Eckelson and I go out hunting sometimes and get a few geese or ducks. The great excitement of the week has been the arrival of the *Nellie Peck* bound for Ft. Benton 1,200 miles up the river. She brought no news for us but will soon be followed by N.P.R.R. [supply] boats.

Missouri River Crossing, May 19

After a long spell of comparative idleness Gen. Rosser has paid us a visit and dispatched Eckelson with a party across the Missouri and started me on my work here. The company acts foolishly it seems to me. Instead of having their lines run ahead of the contractors it is nearly always the case that the contractors arrive on the ground before the line is staked out and then it is hurry-scurry to give them work. This is the case now. I am ordered to stake out work and put men on it before a decent preliminary line even is run through James River. Such operations involve loss to the company and of reputation to engineers and the engineers get all the blame. After Gen. R. left I ran a preliminary line about thirty-five miles from here and re-turned today. I shall locate a small piece of line for the contractors tomorrow and next day, so if they come they can be accommodated <u>instanter</u> [*sic*]. These were Gen. Rosser's orders.

This place is growing fast. Claim shanties are now thick and they have a small town started three miles down the Creek. Still the real bustle has not yet begun as the contractors have not put in an appearance. The Indians made a dash at Fort Rice which you know is on the west side of the River and captured thirty ponies which they got off with. As these ponies belonged to Indian Scouts and not to the Government, I don't attach much importance to the demonstration. I went up Heart River about two weeks ago with Eckelson. I wish you would order me another Maynard [rifle] improved like the one I have. You can send it to Brainerd where I will get it attended to. I want a light gun to carry on horseback and it should have a sling.

Coteaus, June 5

It is fast getting warm and the mosquitoes are voracious, unappeasable, exhaustless, and their name is legion. About seven o'clock they simultaneously rise in clouds from the ground, fall on one, and literally devour him. We build every night a smudge [fire] of buffalo chips. This takes an hour and puts a stop to the outrages of the villains. I believe that the contractors have arrived at the Missouri but do not have many men at work for men are very difficult to get at present. There are so many roads building that there is no difficulty in getting work east, and they are absurdly afraid of Indians. This keeps the company somewhat behind with its work.

I am expecting that gun you are going to send me. It will be very useful as we see plenty of antelope every day but they are shyer then they are across the Missouri and one rarely gets within two hundred yards of them. This I consider a long range particularly with the Winchester rifle or carbine which [is] a company gun I am using. They have now two or three infantry companies at James River and the Sixth Infantry is to rendezvous at Ft. Rice in a short time if they have not already arrived. Eckelson is across the Missouri and has been exploring a cut off route on Hart River to save about nine miles of road. The country is almost mountainous and I do not see any great prospect for success from the attempt to make this deviation.

We are living on short rations. Bacon, apple sauce, and bread being the staples. They do not send supplies forward as they ought somehow and Gen. R. has an idea that because he sent out several hundred pounds of pork, coffee, and flour and sugar that we are all well supplied with rations whereas the ration list provides canned fruit, butter, lard, pickles, etc. Indeed men cannot work well; when they get nothing but salt victuals they are despondent and discontented. I often feel homesick myself. I don' know how it is but now that I ought to be best pleased with my position and prospects I feel more homesick and tired of the country at times than I ever have been before. Perhaps the grub has something to do with my foolish feelings.

Coteaus, June 6

"Mr. Lo"[24] is also in the ascent now it seems and is promising himself some fun this summer and I only hope that he may confine his frolics on this side the Missouri to stealing horses. Every thing out here wears its spring clothes (except the engineers who are as dilapidated a set of mortals as you will see in a long time) and the trees along such of the streams as we are fortunate enough to be blessed with a wear their richest green. A few weeks however will parch and sear this foliage and instead of the sea of verdant green which the prairies now afford there will be the monotonous brown undulations so tiresome to the eye and intellect.

I have just run a line [east] from the Missouri sixty miles and am awaiting supplies from the nearest station at James River some thirty-seven miles distant [from there]. Our fare until the wagons get back is and has been for several days "bread and apples sauce" a fare which long usage has accustomed us to endure without much repining. Speaking of James, I sent a sketch of our camp to *Frank Leslie's Illustrated Newspaper* a few days ago and if they publish it I would be glad to have a copy sent me. Please do up the paper in such a way that it cannot be seen as illustrated papers are in great demand here and are generally gobbled up before reaching their destination.

[No location given] June 14

We have just had a visit from Gen. R who took me with him down to the Crossing. The town there is to be called "Edwinton" and is now

---

[24]"Mr. Lo" was the slang or "cool" expression for an Indian.

partially laid out. I invested $25 to secure one lot on Main Street so you see I am now a landholder. Only two blocks are sold at present and most of those offered have been bought up already. The success of the venture depends upon the RR crossing at the place selected for the town and I think it will go there. I have ventured $25 and if the railroad goes elsewhere I shall not pay up the residue of the $100 asked for the lot. The contractors have arrived and are going to work with a will. They have 150 teams and 100 men at present but expect reinforcements. My line is not yet entirely located, not a fifth even being ready for construction and most of it undecided even where it shall go.

In Camp 40 miles east of the Missouri [Steele, North Dakota], July 2.

Contractors are at work on the west end of my line while I locate eastward. I have the grandest set of scoundrels in the shape of contractors that ever were seen and they give me great trouble. They refuse to finish their work neatly and that alone will get me into trouble with my superior officers. I shall try and get their contract forfeited. Hon. Dr. [Walter A.] Burleigh formerly [a nonvoting] M.C. [member of Congress] from Dakota is their head man of the contract but it is managed by a scamp named H who he pays to "boss the job" as Burleigh does not know the first thing about railroads.

I was summoned to James River the other day. I left the Missouri at 3 A.M. and drove twenty miles by 6:30 A.M. and then with another team drove about sixty miles by ten o'clock the next morning. This brought me within thirty miles of James River so that on a pony I managed after dinner to get in good season. My thirty mile ride was through a continued and violent rain and hail storm accompanied by terrific thunder and lightning. The lightning was so vivid and frequent that I left the road in places to avoid hills where I thought it dangerous to be, and you may imagine how uncomfortable I felt when once a streak of lightning fell not five hundred feet from me and in a deep hollow too, surrounded by hills. The dust and stones flew up in a cloud where the bolt fell and made me think there was no security to be found anywhere.

Coteau Springs 30 Miles west from James River [now Crystal Springs, North Dakota], July 11

I have just come from the west with Gen. Rosser, Mr. [Milnor] Roberts[25] and Mr. Linsley our Chief and assistant Chief Engineers. They arrived at my camp day before yesterday evening and took me with them to the Missouri. They looked over my location on the way west and to my great relief found it satisfactory. Roberts is a dear old gentleman and Gen. R would be if he didn't muddle things up so awfully. He hasn't the slightest method in him, and rushes engineers off their own work into somebody else's and makes confusion all the time. Mr. L, a talented man but a perfect specimen of the hard, cold, calculating Yankee and we are [all] hoping he will soon be shipped [out]. Our trip was on the whole a pleasant one. Mr. Roberts was much pleased with the country and will probably make a report that will give Mr. Jay Cooke great satisfaction. We lived on the NPRR steamer while at the river and went up and down stream for several miles sounding and making preparations for locating the crossing.[26]

Game on the west side of the river is abundant; Antelope, elk and deer both black-tailed and white-tailed. There is now almost no game this side of the river. There is a hunter at the military camp on the west side of the river who takes orders to kill antelope or deer and gets them invariably.[27] He makes sometimes $400 a month—a better business than engineering. I am anxious to see my new Remington and try it. Gen. R has the cheek to propose me to lend it to him to use on the Yellowstone trip. I told him I would very willingly dispose of it to him and order another. But he did not want to buy a new gun, he preferred borrowing. He wanted to borrow my shotgun too and you know yourself how battered a gun gets on such a trip.

[25] W. Milnor Roberts (1810–81) was one of the country's leading civil engineers. A Pennsylvania Quaker (although his extended family included Union and Confederate generals), he had done extensive consulting work for Jay Cooke before the war. Before Cooke invested in the Northern Pacific, he had Roberts lead a small team to investigate the route from Puget Sound, east through Washington Territory, across northern Idaho, and then as far as the eastern side of the Bozeman Pass. Cooke trusted Roberts, and the two became even closer when Roberts became chief engineer.

[26] The steamer, the *Ida Stockdale,* is a good example of Gregory Smith's incomprehensible 1872 purchasing practices. The steamer, a stern-wheeler, was built in the winter of 1866–67 and rated at 364 tons. It was seldom used by the Northern Pacific.

[27] Charles A. Reynolds (1842–76), aka "Lonesome Charlie," died with Custer.

By the way I did not tell you that another expedition is about to leave the Missouri River, July 20, to go through to the Yellowstone. Gen. Rosser and Mr. Linsley are both going and it will be a grand affair. Gen. Rosser is taking all my best men, tents, horses, instruments, etc. from me to send out there. It is a perfect shame, I think, to serve me so yet he treats all the engineers the same. He goes on the "rob Peter to pay Paul" principle. No engineering party here has any outfit. They have just what they can get from each other and patch up as best they can. It is the real southern shiftless management, hurry-scurry, and hop along on one leg method of doing things. It makes me almost sick at times, and yet my prospects are so good that I don't like to leave the road. [However,] if they do not give me the "Bridge across the Missouri," I believe I shall leave soon or else go to contracting. Mr. Roberts is a fine old gentleman and I wish you could meet him. He is genial and kind, a great contrast to Mr. Linsley's bull headedness.

Source: "Letters of Montgomery Meigs," edited by Elizabeth Rodgers, master's thesis, Montana State University–Bozeman, 1937, microfilm 140, Montana Historic Society.

Stanley's Official Report through August 17

David S. Stanley, Headquarters Middle District, Fort Sully, D.T., to Assistant Adjutant General, Department of Dakota, St. Paul, Minnesota, October 28, 1872

Sir:

I have the honor to report that in obedience to Special Orders, No. 142 of July 17th, 1872, and other instructions from Headquarters Department of Dakota. I left this post on the 18th day of July past for Fort Rice, D.T., where I was to organize and fit out a force to escort the Engineers of the Northern Pacific Railroad in their survey of the proposed Railroad route from the mouth of Heart River D.T., to the mouth of Powder River on the Yellowstone, Montana. The escort was to leave Fort Rice on the 22nd day of July, but upon consultation with the Assistant chief Engineer, D. C. Linsley, and Division Engineer, T. L. Rosser, I found that neither myself nor the Engineers could start on that date: The six companies of the 8th Infantry had not arrived; the beef cattle for the Expedition were some days behind; hard bread for the troops was wanting, and upon calculation the transportation provided by the Quartermaster's Department was insufficient.

As soon as possible I consulted with the above named Engineers at Fort Rice regarding their intentions and wishes. They told me they hoped to go at once to the mouth of Powder River, and thence work slowly back to the Missouri, it was also hoped that I could furnish an escort for Mr. Linsley back to the Missouri immediately after reaching Powder River. Assuming the distance from Fort Rice to Powder River to be 300 miles, 20 days must be allowed Infantry to march it, having in charge a large train and especially over this country, most [of it] unexplored, truly, but reported to be very rough and difficult in some portions even for an Indian lodge trail. It could not be expected that a supply train could be sent back, and after reloading at Fort Rice could meet us sooner than 60 days. The grass was already burning east of the Missouri, and it was not prudent to start with less than five pounds of grain per animal per day. In my instructions I was authorized to draw up to the posts from Fort Randall to Fort McKean, for transportation, short. I therefore ordered the A.A.Q.M. of the Expedition to hire 32 two-mule or two-horse teams and wagons, believing that the Department commander would expense rather than to have the Expedition either delayed or started out inadequately supplied with real necessities. One-half of these wagons were hired at the same rates and on the same conditions given Mr. Wilder, the Minnesota contractor; this might have been improved upon, as I employed the remaining wagons, teams and drivers at $5.00 per day per wagon team and driver.

The force comprising the Escort was composed as follows:

Headquarters.

Colonel D. S. Stanley, 22nd Infantry, commanding Expedition.

1st Lieut. H. H. Ketchum, Adjutant & 22nd Infantry, Acting Assistant Adjutant General.

1st Lieut. N. D. Badger, 22nd Infantry, Acting Assistant Quartermaster and Acting Commissary of Subsistence.

Surgeon J. P. Wright U. S. Army, Chief Medical Officer [and Stanley's brother-in-law].

Two Regd. Noncom staff officers, 22nd Infantry and one Hospital steward U.S. Army. Battalion, 8th Infantry.

Companies "A, B, C, F, H and K," 16 officers and 264 enlisted men, Lieut. Col. H. D. Wallen, 8th Infantry, Commanding.

Assistant surgeon H. R. Tilton, U.S. Army, attached to Battalion as Medical Officer. Battalion, 17th Infantry.

Companies "A, C, and F," 7 officers and 119 enlisted men, Major R. E. A. Crofton, 17th Infantry, commanding.

Acting Assistant Surgeon S. J. Allen, U.S. Army attached to Battalion as Medical Officer. Battalion, 22nd Infantry.

Companies "D and G," 4 officers and 89 enlisted men, Captain C. A. Webb, 22nd Infantry, commanding.

Company "F," 22nd Infantry.

3 officers and 65 enlisted men manning 1 12 pounder Napoleon gun, 1 1 inch Gatling Gun and 2 ½ inch Gatling Guns, and performing the pioneer duty of the Expedition, Captain A. H. Goodloe, 22nd Infantry, commanding.

13 enlisted Indian Scouts, 4 Privates detailed from Company "F" 22nd Infantry, and 5 half-breeds employed in the Quartermaster's Department, 2nd Lieut. D. H. Brush, 14th Infantry, commanding.

Total: 33 officers and 553 enlisted men.

I employed as principal guide and interpreter Mr. Basil Clement,[28] a very thorough and enterprising guide, one of the few left remaining of the old trappers of the Rocky Mountains. I employed in addition five half-breeds, daring young fellows, whose services were very valuable.

The allowance of baggage for men and officers was put at the lowest limit consistent with reason. The Commissary was ordered to carry nothing extra for officers, excepting lard, baking powders, and a small supply of ham. As we expected to be out until the 31st of October, I did not think it prudent to dispense with tents.

The order of march and encampment as substantially as follows:

---

[28]Basil Clement or Claymore (1824–1910) was one of the more experienced scouts in the Northern Plains and had been an assistant to the legendary Jim Bridger as a scout and hunter. He was on numerous prewar army mapping explorations and was the lead scout in General Sully's campaign against the Sioux. Although he had been the lead scout on the 1871 Northern Pacific survey, Rosser neither trusted his judgment nor personally liked him, perhaps because of his drinking.

The train moved habitually in two lines; the Infantry was divided into four battalions, one battalion marched as advance guard, two battalions marched, one on each side and opposite to the front and flanks; 12 pounder gun in center of train, Gatlings, front and rear. Coming into camp, a front of camp being indicated, the train formed [a] corral, leaving a space which was filled by company wagons after unloading; each battalion formed line 45 paces outside of the corral and in the same relative position they had marched during the day. Before breaking ranks, the pickets varying in strength according to circumstances were told off and marched under an officer to their immediate front, to the most commanding ground, and remained there until dark, resuming their stations at Reveille, and joined their respective battalions as they moved off in the morning. Night guard was set at dusk, sentinels close in, all animals were brought into the corral, making sentinels beats very short. Reveille under arms invariably just before daybreak. With these precautions very necessary in the Indian country, a surprise was impossible, nor could the Indians at any time run in on our stock without receiving ugly volleys on their first appearance.

Everything was in readiness and the Expedition set out on the morning of July 26th. Six-mule teams were loaded up to an average of 3600 pounds, and two-horse teams to 2800 pounds each. The night of the 26th of July it rained heavily, making the roads very heavy and progress slow. The Engineers and their train joined us the 2nd day out at Cache Le Maise, 20 miles west of Fort Rice, to which point they had an escort from Fort McKean. Pursuing a westward course we crossed the Heart River to the North side, 53 miles from Fort Rice, passed the base of Heart Butte 65 miles from the same post, and reached and re-crossed Heart River to the south bank of the 2nd of August, at the point on the map marked, "Sully's Crossing," having up to the point followed mainly the trail followed by Major Whistler last year.

At the point 106 miles west of Fort Rice, D.T., the trail of last year was quit, and the Engineers commenced their work, designed to run a preliminary line, on a course as nearly as possible due west to the mouth of Powder River. As the line of the Engineers and the wagon road sometimes separated several miles, one company was detailed daily to guard the working party. This company was sometimes strengthened to 50 men and when the neighborhood was thought dangerous the entire force was held near the surveyors during the day. The course pursued kept us south of Heart River, crossing at convenient distances for water, the branches of that stream. On the 5th of August we reached the "Bad Lands" of the Little Missouri. After one day's exploration accompanied by all the Mounted force, Mr. Rosser found a practicable route through the "Bad Lands" 25 miles south of his passage last year.

On the evening of the 7th of August we camped within one mile of the Little Missouri which at this point flows from a west point to the east; following up the right bank of this stream after one day's rest we crossed upon the 11th of August with some little difficulty, the stream being up with rains, and came the same evening upon a small branch of the Little Missouri, named Pettit's Creek, in memory of Mr. G. D. Pettit; a young man in charge of Mr. Wilder's train, who died during the night. He had left home in an advanced stage of consumption hoping to be benefitted. His hope proved delusive, and all we could do for him was done; to nurse him tenderly, and give him Christian burial.

Ten miles west of Pettit's Creek we crossed near its source the large west branch of the Little Missouri, called on railroad map of last year "Inman's Fork"[29]—called on Sully's map its Indian name, Wak-pas-chah-pah, or Beaver River. I recommend that the latter name be retained on the official map to avoid confusion. Five miles after passing this stream we were suddenly brought up by "Bad Lands" stretching off many miles right and left, and more difficult at sight than those of the Little Missouri. We camped the evening of the 12th on the crest of these "Bad Lands" and here first saw Indians, about 30 in number, watching us. The next day we made the descent to the creek running through these Bad Lands, which was named Cabin Creek from some old Indian log cabins found standing near our camp. Leaving the valley of this creek and its "Bad Lands" we found water next day running due west, and on the 15th found a large running creek, course North west, wooded bottom, ¾ mile wide, which was mistaken by some for Powder River. We

[29]Inman's Fork had been named by Rosser for the equally able and jovial Capt. Henry Inman (1827–99), everyone's friend and always in debt. Sadly, Inman was transferred into the quartermaster corps where he sold two army horses and kept the money. By 1872 he had been cashiered from the service. Gen. John Pope overturned a court-martial that declared him innocent, and his name was removed from all maps. Later Inman became a best-selling author, but died destitute.

descended this creek 5 miles the afternoon of the 15th, and camped as far from the cottonwood thickets as we could get with any convenience.

The next morning just before Reveille and just before daybreak, a party of probably twenty Indians rushed out of one of the clumps of wood, yelled, fired a volley into camp, three or four riding very close to our sentinels, and ran away as fast as their ponies could go. One and a half days' march down this stream, which proved to be O'Fallon's fork, brought us to the Yellowstone. Our scouts continued to observe Indians following us, and just as we reached the Yellowstone one of the Scouts ran in with the information that a very large number were coming fast towards us just behind the hills. In a few minutes everything was ready for them, but they made no further demonstration. We camped on the Yellowstone the night of the 17th.

Source: "Report on the Yellowstone Expedition of 1872," No. 917.86 Sta., Montana Historical Society, Helena.

Newspaper Accounts

The following newspaper and journal accounts offer various contemporary views, insights, and prejudices regarding the eastern survey from both military and Northern Pacific viewpoints. While accuracy is sometimes wanting—one article stating 20,000 warriors were arrayed against the expeditions—taken as a whole the articles give the reader and researcher an unmatched contemporary flavor. In many cases the articles are repetitious or too long to print, and I have made what I hope are appropriate cuts.

The name of the newspaper and date should be considered the source of each article. And, as often as not, the publishing date is weeks after the story was written—reflecting that written communications had to be carried by hand to a location with telegraphic or postal service. The reader is also reminded that until early 1873, "Bismarck" was called Edwinton, the Crossing, the Northern Pacific Crossing, or Carleton.

Sioux City Journal, May 24

TRIP OF THE *MINER*
FROM [FORT] RICE TO SIOUX CITY—
A RAID ON GOVERNMENT STOCK—
THE CROSSING OF THE NORTHERN PACIFIC—
FIRED ON BY INDIANS, ETC.

JIM EMMONS[30] OF THE STEAMER *MINER*
FURNISHES US WITH THE FOLLOWING "NEWSY"

MAY 16—The steamer *Miner* arrived at Fort Rice at 8 A.M.—has freight discharged by noon, but owing to the prevailing high wind, laid to until 6 P.M. The noble Hunkpapas made a raid upon the Government herd of ponies and mules at Fort Rice and captured thirty head. The Government [Indian] scouts are now in pursuit of them. The thanks of boatmen, particularly the crew of the *Miner*, are due to Captain Charles E. Clark [acting] commandant at Rice[31] and to Lt. Chance, officer of the day, for courtesies extended. Mr. A. O. Eckelson, surveyor of the west end of the Northern Pacific Railroad, has arrived at Rice. His party came through from Red River without molestation and were to proceed to Heart River on the 20th to resume the surveys of the road. Captain Gassman, a picked officer, will have command of the escort. It is probable that the party will encounter some Indians.

The embryo town of Carlton has of yet no permanent location. Some 200 teamsters and woodchoppers have located near this point on the east side of the river and have staked claims, but a good many of them are in destitute circumstances, if work does not commence soon some of them will have to "launch their little barks" and float downstream. From a friend I learn a few of the market quotations. Flour is selling at $5.10; whiskey, steady at 50 cents; truth flat and not in demand. The officers

---

[30]James A. Emmons (1843—1919) was a former steamboat captain and in 1872 a part-time Northern Pacific employee stationed in Edwinton/Bismarck. He was responsible for all NP freight traffic, co-owned a number of related transportation businesses, and was one of Bismarck's founders. A county in south-central North Dakota is named for him.

[31]Charles Edward Clarke, not Clark (d. 1901). Little is known about Clarke other than that in 1868, having been informed by a post surgeon that a black enlistee, Cathay Williams, was actually a woman, he had her discharged from the army. Clarke is known to have been on the 1871 and '73 Yellowstone surveys.

at Fort Rice seem to think that there is nothing definite as to where the crossing will be until the company commences to drive their piles.

Army and Navy Journal, June 1

LT. GEN. PHILIP H. SHERIDAN REPORTS ON HOSTILE INDIANS

MILITARY DIVISION OF THE MISSOURI—General Sheridan forwards to the War Department official reports dated Fort Stevenson, April 19, saying that from information derived from "Lone Feather," "Sharp Horn," and other chiefs and warriors of the Arickera tribe,[32] with whom the Indians hostile to the whites are on terms of intimacy, it is learned that there are now collected at a point lying southwest of Fort Berthold, and distant from that point, about 120 miles, about 2,000 hostile Indians, composed chiefly of members of the Hunkpapas, Minneconjous, Ogallalas, Sans Arcs, Brulés, Blackfeet, and Cabheads,[33] who have declared their intention to oppose the progress of work on the Northern Pacific Railroad this summer. They have torn up the stakes planted by the surveying party, which marks the route of the road in the Powder River country.

The above mentioned tribes have been engaged in war during the past with the Crows and Nez Perce, in which the former lost a great number of horses. This has rendered them desperate, and they have expressed their determination to obtain horses to replace those so lost, and to this end from about the 1st of May small parties will be detached from the main body to endeavor to capture horses and other property from Forts Buford, Stevenson, Rice, and Totten, and from the sutlers and others in the neighborhood of Apple Creek.[34]

When the railroad company shall have pushed their work to the west side of the Missouri, these Indians contemplate a concentration of their forces for a general hostile movement en masse against the company. They number among their leaders the following well known hostile Indian chiefs: Sitting Bull [Hunkpapa], Black Moon [Minneconjou], Iron Dog [Two Kettle], Iron Horn [Hunkpapa], Four Horns [Hunkpapa], and Long Dog [Hunkpapa]. These Indians are all well armed and equipped.

[32]Spelling has been modernized for this and other Indian tribes.

[33]The first six tribes identified in this sentence were Teton or Lakota Sioux. The seventh tribe, the Two Kettles, are not mentioned unless that was the tribe indicated by the word "Cabheads."

[34]Where the Northern Pacific would cross the Missouri.

Minneapolis Tribune, June 12

LETTER FROM ENGINEER ROSSER

NO TROUBLE WHATEVER ANTICIPATED

The recent atrocious murders committed at Oak Lake [Minnesota][35] by a set of straggling Indians and one or two other acts of barbarity perpetrated by some drunken savage at widely separated points on the frontier—have caused quite current rumor that the Indians in general were contemplating a fierce war upon the whites wherever they should find them defenseless, and where there was rich promise of booty.

Especially of late it has been noised abroad that on the line of the Northern Pacific Railroad the Indians maintained a belligerent attitude towards the advance force of Engineers, and were causing them no end of troubles and alarm. Beyond question these reports are groundless and emanate from parties hostile to the interests of the road. The following letter from Gen. Rosser, who is Division Engineer of the Dakota Division, disposes of the matter and should serve to quiet the apprehensions which many folk have felt. It bears the date, Fargo, Dakota, June 9th.

Hon. D. Morrison[36]

Dear Sir—

Rumors which have evidently been set afloat by designing men, concerning the hostile attitude of the Indians in Dakota, have been so generally circulated that I feel it due not only to the interest of the emigration and the progress of the N.P.R.R., but to the Indians themselves, to arrest the further circulation of these false reports.

In the first place there are but few Indians east of the Missouri River, and amongst them I have lived for nine months, and have made regular trips once a month from Fargo to the Missouri River without an [military] escort, and have very rarely met Indians, but all that I have met have shown a most friendly disposition, and I have never found it necessary to guard my horses or supplies from them. I have small [surveying]

[35]Oak Lake or Oak Lakes Cut was a rail stop established in 1871 by the Northern Pacific Railroad in west-central Minnesota near the town of Audubon. Today the site is a ghost town.

[36]Dorilus Morrison (1814–97) was a Minnesota legislator, two-term mayor of Minneapolis, and a leading banker and businessman responsible for much of Minneapolis's early growth.

parties strewed over the [planned] line from here to the Missouri, and one party 40 miles west of the Missouri and none have been troubled by Indians. The party west of the Missouri has a small escort, but those east of the Missouri have none.

Most truly yours . . . [etc.]

Sioux City Journal, July 12

DAKOTA: FATAL DISASTER AT FORT RICE
SPECIAL DISPATCH TO THE JOURNAL

FORT SULLY, JULY 11—A small flat boat, used as a ferry across the Missouri River at Fort Rice, sank three or four days since, with four men belonging to the Seventeenth Infantry, and a quantity of baggage belonging to Capt. [Clarence E.] Bennett of that regiment. Two of the men swam ashore, but the other two clung to the boat and were drowned.[37] The bodies not recovered. The boat was of but little value, but its loss will be severely felt, as there is now no way to get across the river at that post. Capt. Bennett lost about $2,500 in money and valuable papers.

The following article is a fine example of garbling facts, unintentionally or not.

Sioux City Journal, July 16

FROM UP RIVER
THE NORTHERN PACIFIC RAILROAD— A FORMIDABLE MILITARY ESCORT FOR THE PURPOSE OF EXTENDING THE SURVEY— AN INDIAN MURDER, ETC., ETC.

Fort Sully, D.T., July 6, 1872

Dear Journal:

On my arrival here this forenoon I learned that an expedition of fourteen companies, twelve infantry and two cavalry, under General Stanley,

[37] How one's memory plays tricks! Twenty years later, one of the people who possibly witnessed the incident, Linda Slaughter, the wife of Fort Rice's surgeon, correctly recalled the regiment involved, but mistakenly thought that over a dozen soldiers had drowned, not one person surviving.

has been ordered as an escort to the corps of engineers who leave Fort Rice on the 30th inst., to extend the survey of the Northern Pacific Railroad.[38] There will be six companies from the Eighth Infantry . . . [list of officers omitted]. The latter battalion will be under command of Major Crofton. Two companies of the 7th Cavalry are expected to accompany the expedition and also a battery of Gatling guns.

The expedition will march from [Fort] Rice to the Black Hills,[39] thence to Powder River and from there to the Yellowstone and back home.

The following officers compose General Stanley staff . . . [names omitted].

A soldier of the Cheyenne Agency was murdered and scalped on Wednesday morning last [July 3] just outside of the agency by an Indian.[40] The Indian lit out for the hostile camp immediately upon commission of the act.

The river has raised two feet here within the past twelve hours and still is rising. The river is full of floating logs.

[Signed] D.S.[41]

St. Paul Weekly Pioneer, July 19

NORTHERN PACIFIC RAILROAD
TRANSIT AND TOPOGRAPHICAL SURVEY
WEST OF THE MISSOURI

We understand that Gov. Smith, President of the Northern Pacific Railroad, has recently ordered a new and very critical transit and topographical survey of the region of country west of the Missouri river, and the survey to be conducted under the direction of Gen. Thos. L. Rosser, the distinguished Chief Engineer of the Dakota Division. Ample arrangements have been made with the military authorities for a large and effective escort to accompany the observers. The corps of engineers are being made up, preparatory to take the field on the 20th inst. It is proposed to extend examinations of the country even beyond the Yellowstone River,

[38]The forces are overstated, least of all any cavalry.

[39]This is also totally erroneous, but would have been of considerable interest to those who (correctly) believed that gold was to be found in the Black Hills.

[40]The agency or reservation was on the western side of the Missouri where the Cheyenne River enters it, some thirty miles north of Fort Sully, South Dakota. The site is now covered by Lake Oahe.

[41]Not identified, but not Colonel Stanley.

to a connection with examinations of the country coming from Central Montana. When this survey is completed, a full and critical examination will have been made from Duluth to Puget Sound. Doubtless this extensive survey about to be put on foot, going both east and west, is being done with a view to the extension of the road at least from the Missouri river to the Yellow Stone. This survey will be completed, and the notes reduced to tabular statements, showing sufficient data and preliminary facts upon which to invite, and even accept, if necessary, propositions for construction by the 1st of November next.

Sioux City Journal, July 21

FROM FORT BUFORD
ECCENTRIC WEATHER—TROUBLESOME INDIANS—BOAT RACING—MOSQUITOES' PARADISE—BOAT NEWS

Steamer *Western*, Fort Buford, D.T.,[42] July 10, 1872

Dear Journal:

We cleared for Sioux City this evening with a snow storm in feeling distance of us. Two days ago the heat was oppressive here, the thermometer was doing business at 108 degrees in the shade, but this evening the mercury has gone down to 30 degrees. The new garrison of soldiers here think they have got into an eccentric climate. About next January I reckon they will be positive on that point.

We reached here at 6 o'clock this morning, fifteen days and twelve hours out from Sioux City, which is remarkably good time when we take into consideration that our boat has handled over 500 tons of freight on the trip, and 400 tons of that was way freight, which we received and discharged at every fort and agency on the river between this place and Sioux City. In addition to these sources of detention, we had on board

[42]The *Western* was a 400-ton stern-wheeler built in Pittsburgh in 1870–71 but poorly designed for the upper Missouri trade. Fort Buford was located in present-day northwest North Dakota just east of where the Yellowstone River enters the Missouri. It was named for John Buford (1826–63), West Point '48, the highly regarded cavalry general who temporarily blocked Lee's forces as they neared Gettysburg on July 1, thus permitting Meade's army to gain the higher ground. He died of typhoid fever.

Major Smith, the new U.S. Paymaster, who paid the troops at Sully, Grand River, Rice, Stevenson, and other points.

The Indians are still on the warpath and their depredations are growing more numerous as the season advances. Many of the friendly Indians are breaking loose and putting out for the hostile camp. The Cheyenne Agency is furnishing its full quota of these "good Indians." Three left there for Sitting Bull's camp last week. At Grand River they attempted to murder a post herder by the name of Rummer, within forty rods of the Agency buildings. The herder was shot through the arm, but made his retreat into the post with his hair.

Col. O'Conner, of the Grand River Agency, sent a delegation of Sioux Chiefs by the steamer Sioux City to Fort Peck [Montana], to meet the old hostile chief, Sitting Bull. It is hoped that they may be able to negotiate terms of peace with him.

About two weeks ago a party of Rees Indians captured and killed three Sioux squaws and five children on Bean Creek. The same party fired at a Mackinaw near Jim Gayton's woodyard, but did the occupants of the boat no injury. A party of Rees Indians left Stevenson a few days ago and captured six Sioux squaws about six miles from there. They killed and scalped them all and then returned to Stevenson, where they displayed the bloody trophies of their cowardice and had a scalp dance.

We reached Carlton, or Burleigh, or Edmondstown [*sic*], or Wellington, or the Crossing, just as you like it, on the 17th, very early in the morning. That town or bouquet of towns, is the paradise of mosquitoes. The most industrious, savage, and blood-thirsty mosquitoes I ever saw are there. Nobody pretends to sleep; when any one of the citizens becomes exhausted he builds a "smudge" in the center of his bed chamber, or place of business, and chokes himself into unconsciousness with the smoke. We slept not a wink on the *Western* for forty-eight hours while running through that country. There was no place on the boat that was secure from the pests. I "cached" myself under a tarpaulin, wrapped an immense rubber blanket [poncho] about me, put on gloves and overshoes, and tangled myself up nigh unto suffocation, but was no use. I was meat for them all the while.

Philadelphia Enquirer, August 6

LETTER FROM CHIEF ENGINEER ROBERTS
RAILROAD TRAFFIC—CHARACTER OF DAKOTA LANDS—
AN EXCELLENT COUNTRY—INDIAN MATTERS

Chief Engineer W. Milnor Roberts, who has recently (July 15, 1872) completed a tour of inspection along the line of the Northern Pacific Railroad in Minnesota and as far west as the valley of Heart River in Western Dakota, writes as follows of the results of his professional observations:[43]

> The newspaper stories about Indians and danger from them to the railroad across Dakota, have no solid foundation in fact. The Indian situation, so far as the Northern Pacific railroad is concerned, is this: As far west as the Missouri River the question is settled, and no danger need be apprehended. Our men travel all over the country between the Red and Missouri rivers by day and by night, without escort and unarmed. The few troops at the James and Missouri rivers will suffice to prevent thieving raids by straggling natives, until the last have finally disappeared before the army of white settlers. West of Fort Ellis, in Montana all the way to the Pacific coast, the Indian question is forever disposed of. All our engineering parties, six in number, between Montana and the Pacific are operating without escort.
>
> The only portion of the line yet in abeyance between the Indians and the whites is between the Missouri river and the westerly bend of the Yellowstone [Livingston, Montana], east of Fort Ellis; there are parties, now taking the field, [that] have strong escorts of United States troops, well prepared to fight, if necessary. *But no fighting is to be apprehended* [emphasis added]. If our escorts were weak, there might be some risk. Frontier speculators who wish to see government money freely spent in their vicinity and to their advantage will doubtless continue to furnish the newspapers with prophesies of imminent "Indian Wars" in the Northwest.

[43]The Northern Pacific sent this letter—whether Roberts actually wrote it is unknown—to dozens of newspapers around the country. The *New York Times*, for example, printed the letter the following week. The letter's clear purpose was to reassure investors that Indians would not and could not endanger construction. Sections not relating to Indian hostilities have been omitted. Ironically, the letter began to appear in the newspapers just as fighting began.

The construction of the Northern Pacific railroad will entirely settle the Indian troubles on a belt at least one hundred miles wide and fifteen hundred miles long, embracing one hundred and fifty thousand square miles, a large portion of which will ere long contain an industrious population, giving profitable employment not only to the main line but to the numerous branch feeders which will surely be opened as the country becomes settled and its many resources developed.

Yours . . .

No city was impacted more by the Northern Pacific's plans than Sioux City, which knew it would lose its Upper Missouri steamboat traffic primacy. Thus the Sioux City Journal *happily printed negative stories, rumors, or just plain dribble about the Northern Pacific. The following paragraph—its writer unidentified—was factually impossible but nevertheless repeated around the country.*

Sioux City Journal, August 13

ON THE RIVER
FORT MCKEEN AND THE CROSSING, THE NEW TOWN OF EDWINTON

At [Fort] Rice I learned that the Yellowstone expedition has been attacked by two columns of ten thousand Indians each.[44] The news was brought in by two Indian scouts. The report is pretty generally credited by up-river men, as there are but few Indians remaining on the reservations along the [Missouri] river. In addition to this fact, the Indians have been manifesting decided symptoms of hostility all summer, a sure indication of war; besides that, they have frequently and openly declared their intention to prevent the [surveying] expedition from marching through their country. I would not be surprised to [shortly] learn of the confirmation of the report.

---

[44]In Kingsley M. Bray's "Teton Sioux Population History," he estimates that the seven Lakota (Teton) Sioux tribes had a *combined* (men, women, and children) 1870 population of 14,370 of which perhaps 20 percent were warriors. Similarly, Lakota allies such as the Northern Cheyenne, Arapaho, and the remnants of the Eastern Sioux tribes could barely have mustered an additional 1,000 warriors. While contemporaries might have estimated slightly higher figures or even an increased order-of-magnitude, the number 20,000 was absurd on its face.

Western Survey

Maj. Eugene M. Baker and Maj. John W. Barlow

Col. David Stanley's western counterpart was Maj. Eugene Mortimer Baker (1837–84), West Point, '59, who came to Sheridan's favorable attention leading the First U.S. Cavalry in the Shenandoah valley (1864). Had Baker led a state regiment and not a federal one, he undoubtedly would have become a brevetted brigadier general. Baker commanded Fort Ellis, astride the Bozeman Trail, a few miles east of Bozeman and almost exactly where William Clark had camped in 1806 before crossing into the Yellowstone valley.

In late 1869 Sheridan selected Baker to lead a winter expedition against a Blackfeet tribe. Sheridan wrote Baker's commander to "Tell Baker to strike them hard," but Baker struck the wrong village, likely because of his guide's error. Without a single casualty, he murdered or let die 173 Indians—mostly women, children, and old men.[45] When news leaked out, it became a national issue correctly earning Baker a notorious reputation.

Fair-haired, Baker was intelligent and aggressive, physically dominating and a natural leader apparently well liked by most of his officers and men, to say nothing of Montana's public. But he was also an alcoholic, by 1872 no longer able to control his drinking. Chief Engineer Milnor Roberts spent almost a month with him in 1871, but never informed President Smith or Jay Cooke. This failure to press for Baker's replacement—or, if Roberts had, not being emphatic enough—would haunt the surveyors in the months ahead and be the first of 1872's great what-ifs.

[45]The number remains disputed, but numerous sources give similar figures. See Thrapp's *Encyclopedia of Frontier Biography* for the 173 figure.

Baker's command included 187 officers and men from four companies of the Second Cavalry and 189 from four Seventh Infantry companies, their supplies packed in sixty-five wagons. The number of wagons was half that of the eastern survey as the Yellowstone valley provided lush grass for animals. The surveyors numbered perhaps twenty and had three wagons. A handful of herders watched over the cattle. Additionally, perhaps twenty heavily armed prospectors joined the survey, hoping to safely pan for gold. Counting everyone, there were perhaps 500 men, and 700 horses, mules and steers.[46]

Joining Baker was Sheridan's personal representative, the devout Maj. John Whitney Barlow (1838–1914), West Point '61. Barlow had a distinguished war record, was Sheridan's chief engineer, and had joined Ferdinand Hayden's 1871 exploration of Yellowstone Park. Barlow's report was to be printed by Congress, and no other 1872 document is its equal.

Barlow's narration is not just thorough and literate; the reader should note its highly nuanced description of Baker. In discreetly coded language he gives a comprehensive indictment of Baker's leadership. Barlow's review of August 14's fighting is singularly instructive; while other officers are mentioned for gallantry, the reader will search in vain for Baker, an omission of damning importance. In 1872, no one could possibly miss this, for Baker was responsible for leading his men in battle. In addition, Barlow's depiction of the survey's termination before reaching the Powder River becomes the key to understanding the expedition's failure.

Barlow's Journal, July 31–August 13

Secretary of War to the 42d Congress, 3d Session, Senate, Ex. Doc, No. 16

The report of Major J. W. Barlow, who accompanied a surveying party of the Northern Pacific Railroad, in relation to Indian interference with that road.[47]

---

[46]My 700 number: cavalry and infantry officers, 200+ horses; 68 wagons averaging four mules (although many were six-mule wagons), 300+ mules, and 200+ head of cattle.

[47]Formal forwarding letters omitted. Italics in original. Where initials are used for a first name, the name is now spelled in full.

Under verbal instructions from the Lieutenant-General [Sheridan], a journal of the route was kept by me, a copy of which is herewith submitted for his information.

Very respectfully, your obedient servant,

J. W. BARLOW
Major of Engineers.

JOURNAL

The surveying party was under the direction of Mr. John A. Haydon,[48] and numbered about twenty men, consisting of Mr. Haydon, chief of party, two principal assistants, in charge of the compass and level, together with rodmen, chainmen, and others. The escort, commanded by Major Eugene M. Baker, Second Cavalry, consisted of four companies, one hundred and eighty-seven men, of the Second Cavalry, and four companies, one hundred and eighty-nine men, of the Seventh Infantry. The aggregate of troops, surveyors, teamsters, &c., was about four hundred.[49] The engineers' supplies were carried in three four-mule wagons. Sixty-five wagons, ambulances, &c., accompanied the escort. Rations for one hundred and five days for the men, and thirty days' forage for the horses, were taken along.

Preparations for the movement were completed on the 26th of July, 1872, and on the next day the infantry and wagon train left Fort Ellis, the cavalry overtaking them on the 28th.

The route from Fort Ellis to the Yellowstone was through the "Bozeman Pass," and along the line surveyed last year by the Northern Pacific Railroad engineers. This part of the route is extremely picturesque. After crossing the summit of the "divide," the road follows the valley of Trail Creek, to the Yellowstone Valley, where, from various points, the scenery is grand and beautiful. Toward the east the Yellowstone is seen, as it leaves the mountains and plunges downward into the plain below, while

[48]John A. Haydon (ca. 1827–?) apparently had been a competent Confederate (from Tennessee) engineer and captain during the war. He was a last-minute Northern Pacific hire, but other than his being married, nothing is known about him.

[49]Barlow's number should be 500. Based on his own numbers of 376 officers and men, some staff including himself, surgeons, and scouts (at least fifteen total), twenty surveyors, sixty-five teamsters (although some were possibly army personnel), and twenty or so prospectors, the number reaches 496. Sheridan also uses the number 500 in a September 10 telegram.

above, and beyond, the vast billows of lofty mountain-peaks, crowned with perpetual snow, stretch far away to the south and east.

To the north of the river, thirty miles distant, and towering amid the clouds, is seen the ragged, precipitous range known as Crazy Mountains, an admirable name for this wonderfully tumbled wilderness of gigantic rocks.

On entering the Yellowstone Valley, which is broad as well as beautiful, we find that the foot-hills rise from the bottom-grounds in successive terraces, forty to fifty feet in height, as regular and well-defined as the surfaces of a fortification, each terrace indicating as clearly as possible the different water-levels in former times.

The river this season was unusually high, its depth being eighteen feet, and its width about one hundred yards, at the Mission ferry.

Crossing, even with a skiff, is attended with great difficulty, owing to the swiftness of the current, which here runs at least eight miles an hour.

The lower terraces in this part of the valley are composed of local drift, though in places very-fertile, particularly along the bottoms of tributary streams, as at the Crow Indian agency, where fine crops of vegetables are raised with but moderate labor.

This agency is located upon a bluff south of the Yellowstone, and nearly opposite the mouth of Shields River. It is subject to severe wind-storms, which render the site extremely disagreeable.

The Crow Indians were encamped near the agency, and an invitation was given by Colonel Baker for some of their young men to accompany the expedition in the capacity of scouts. None, however, accepted. They stated that they would like to go and fight the Sioux, but not to travel peacefully through their country, and also intimated that we would find bad friends down there.

Mitch Boyer,[50] a half-breed Sioux (though living among the Crows at their agency), was engaged as guide to the expedition. On the 30th of July we went into camp on Shields River, about one mile from its mouth and thirty-three miles from Fort Ellis. The valley of this river is fertile and well adapted to farming purposes. Irrigation from the river is, no doubt, practicable for many miles. The uplands through this entire region are perhaps more fertile than the river-bottoms, for the soil has

[50] There are numerous spellings of his last name, including Bouyer and Bowyer.

a less proportion of gravel, and if irrigation can be successfully applied to them they will eventually be considered the most valuable lands in the Yellowstone Valley.

To avoid the precipitous bluffs at whose base the Yellowstone flows just below the mouth of Shields River, the heavy train was compelled to make a detour.

JULY 31—We encamped on a warm creek, about two miles from the Yellowstone, after a fatiguing march among the foot-hills of Sheep-Head Mountain. The latter is a spur of Crazy Mountain, and its face to the south is a nearly vertical wall of basalt, extending for miles along the direction of the Yellowstone River, the intermediate country having seemingly subsided from a much greater altitude, and been subsequently washed into gullies with intervening ridges.

A great variety in the rock formation was observed in this region. Gray sandstone of a fine variety is abundant, also beds of limestone with traces of lignite and gypsum.

AUGUST 1.—The command continued over a region similar, though less rugged than that traversed the previous day, and encamped in the Yellowstone Valley, one mile from the river, on a cool mountain stream known as Cottonwood Creek.

The "Crazy Mountains" are still in plain view, and seem to be a continuation of the snow-range south of the Yellowstone, the similarity in general appearance being very striking. The character of the soil in this portion of the valley is extremely rich, and a fine crop of bunch-grass covers the entire region, though there is a marked scarcity of timber.

A growth of stunted pines is seen in the ravines of the distant uplands, and a fringe of cottonwood is usually found along the borders of the streams.

The camping-ground on Cottonwood Creek combined every advantage, being level and having wood, water, and grass in abundance.

The weather was warm, the thermometer standing at 88° in the shade, and in the evening the mosquitoes were somewhat troublesome.

AUGUST 2.—The command moved along the nearly level bottom of the Yellowstone for about seven miles, and camped on a considerable creek bordered with a heavy growth of cottonwood, nearly all of which, however, is now dead and affords excellent fuel. A good deal of delay was caused in crossing a sharp ravine, picks and shovels being required to improve the roadway. The temperature increased, the thermometer reaching 90° in the shade. At this point the Yellowstone has an exceedingly rapid current, falling at the rate of eleven feet to the mile. It admits of being conducted in canals to some of the higher terraces, and will thus serve to irrigate extensive areas on either side of the stream. The water in the creek on which we encamped was strongly alkaline, and increases, rather than quenches, thirst.

AUGUST 3.—In moving forward to-day it became necessary to cross a broken space about two miles in extent, intersected by ravines and water-courses, and then the level Yellowstone Valley was entered. Good progress was made for five miles, when the command encamped a half mile beyond the Big Timber on the immediate bank of the Yellowstone. All the streams in this section, including the Yellowstone, are alive with trout, yielding a supply for the entire command. The soil remains fertile, the grass-crop being sufficiently abundant to be mowed for nightly feeding.

AUGUST 4.—The river at this point is 300 feet in width, some 12 or 15 feet deep, and booms along in great waves at the rate of seven or eight miles an hour. It would be impracticable to cross this stream with the facilities at our command, and we are therefore compelled to work back among the hills to avoid the high bluffs just below and near the mouth of the Sweet Grass.

For several miles before reaching these hills the road was very little obstructed, but upon entering the rough country masses of sand-rock were found cropping out from the ridges, whose sides are covered with stunted pine trees. Cactus is found in great abundance, and also a miniature growth of sunflower some two feet in height. After a march of nine miles we camped near the Sweet Grass, several miles from its mouth, in a pleasant valley with grass, wood, and water in sufficient quantity. This stream derives its name from an aromatic grass which grows along its banks and which has an exceedingly agreeable and lasting perfume.

AUGUST 5.—The Sweet Grass was crossed this morning soon after leaving camp. Considerable delay occurred on account of the steepness of its banks, which had to be cut down. The stream is shallow and forty yards across. The descent to the stream was so steep that the troops had to let the wagons down gradually by means of ropes. The men were likewise employed in hauling them out on the opposite side.

Winding northwesterly about two miles, we crossed a sluggish stream flowing over a muddy bottom. This is a small tributary of the Sweet Grass, along whose valley our course lay nearly eastward for several days. The grazing in this valley was exceedingly fine, being a rich growth of buffalo-grass. The hill-sides, steep and rocky, are usually covered with stunted spruce pines.

AUGUST 6.—We continued up the valley, crossing and recrossing the small stream at frequent intervals, and gradually ascending to a much greater altitude. This part of the march was thoroughly agreeable; good water, wood, and grass were in abundance. Along the streams cottonwood is found and also wild currants in great profusion.

AUGUST 7.—We left the valley of the Sweet Grass, having ascended to the plains near the summit of the divide between the Yellowstone and Musselshell Rivers.

Our progress thus far had averaged about eight miles per day. After crossing the divide the character of the country materially changes, in appearance at least, the sharp, wooded hills, giving place to an open, rolling country, almost destitute of vegetation, excepting buffalo-grass and wild sunflowers.

The views were extensive and magnificent, though somewhat cheerless and monotonous from lack of vegetable and animal life, Camp was made on the 7th, near a small stream running southeasterly, a tributary of the Yellowstone. A few cotton wood trees were here found, which served for fuel. The guide informed us that a supply would have to be carried forward from this point, as no more fuel would be found for two or more marches.

The same geological formation still exists that has been noticed along our route for several days, viz. fine gray sandstone, in strata, underlying the entire country, and broken up by volcanic agencies, the strata assuming all possible inclinations, being often nearly vertical. Toward evening a severe and cold rain-storm came on accompanied by thunder and lightning, and continued with more or less violence throughout the night.

AUGUST 8.—After a cold and dreary march of four miles, through the rain and over a rolling country, our progress being necessarily slow on account of the wet, heavy roads, we encamped on the brow of a bold bluff overlooking a vast plain to the eastward. Water and a small quantity of wood were found in a deep ravine near camp. Fuel was absolutely necessary in this camp, both for heating and cooking purposes, and was obtained with some labor from the bottom of the ravine.

AUGUST 9.—The scenery along the route to-day was enhanced by the brilliant appearance of Crazy and Yellowstone Mountains, now sparkling in the sunlight with a dazzling coating of snow, received while we were shivering from the cold rain of yesterday. Game is becoming more abundant; herds of antelope are often seen, and the hunters now secure plenty of venison for the command. An occasional buffalo is observed, but no herds as yet. We descended from the elevated table land and entered upon an extensive prairie country, bounded on the south by a low ridge which separates us from the valley of the Yellowstone, and on the north by the face of a precipitous bluff, the side of an extensive plateau reaching off toward the Musselshell River. Between our line of march and this high bluff is a succession of small lakes, plainly visible from different points in our route. These lakes are composed of surface-water, and are said by the guide to be strongly alkaline; the few small streams crossed on this prairie probably become entirely dry later in the season.

A march of eighteen miles was made and camp located on a small stream, which flows but a short distance and then disappears. A few cottonwood trees were found in its valley, and lignite in small quantities was discovered in a lateral ravine. Rain came on again during the night, with violent gusts of wind, which threatened to tear the tents from their fastenings.

AUGUST 10.—The route continued eastward over a rolling country, through numerous prairie-dog towns. Many rattlesnakes were seen, and game is becoming still more abundant; wood and water are scarce, and the grass is getting shorter, hardly sufficient for the stock. Prickly pear is plentiful, and sage-brush is beginning to appear. After a march of only six miles, camp was made on the brink of a sharp, rocky ravine, at the bottom of which was found a fair supply of rain-water, in pools, and a rank growth of willows. The water is probably permanent. This is an important water-course at certain seasons, it being the outlet of an immense territory. It breaks through the ridge southward, forming a wild can[y]on some hundreds of feet in depth, and several miles in length. On the 11th we set out in the hope of making our nest camp on the Yellowstone; the country passed over was open, and for several miles there appeared no great obstacles to the progress of our train, the country being a continuation of the plain of the previous day. At a distance of about seven miles from camp we came upon the brow of a high bluff overlooking a broad terrace beyond. Down this bluff it became necessary to find a practical road for the wagons, and after considerable search a winding route was decided upon, which led down a rough and foreboding-looking spur, which eventually required the lively use of pick and shovel to become passable.

Part way down the descent I noticed a quantity of pure mica, in bright sparkling scales. The soil now became exceedingly poor, without water, though badly cut into sharp, rugged ravines. A few herds of buffalo and numerous bulls, roaming apart from each other, were observed near our route. Antelope were abundant. Owing to the broken condition of the country it was considered impracticable to reach the river before camping, and a suitable locality for this purpose, in this desolate region, became a matter of importance.

The guide reported that water of an inferior quality could be found a few miles farther on and somewhat to the north of our course. Accordingly the proper direction was taken, and at the point indicated a deep gully was found containing a small stream of alkaline water, fed by a few springs. The springs bubbled up through the rock, and were cold and clear, though of an unpleasant taste. The grass at this point had been nearly consumed by the buffalo, and except for the rank growth of prickly-pear the ground would have been nearly destitute of vegetation. Distance traveled eighteen miles.

AUGUST 12.—A slight rain fell last night; camp was broken with the determination of reaching the river to-day. The course taken was about southeast, along the right bank of the ravine on which we had just encamped. Numerous smaller ravines intersect this one, cutting up the country so badly that it was feared our trains could not reach the river in one more march.

The knowledge and skill of our guide were equal to the emergency and he succeeded in finding a route by which the column avoided the gullies almost entirely. The same sandstone formation still continued.

The ravines cut through the strata and thoroughly expose them to view, so that their thickness can be readily measured. In many cases cones and spires of beautiful proportions have been left standing, giving the appearance at a distance of Indian encampments, or the ruins of ancient fortifications. Numerous buffalo were seen along the line of our march, and a large herd was driven through the column, and a general attack made on them by the officers and soldiers.

On reaching the valley of the Yellowstone, some ten miles from camp, a glorious scene was presented. The bluffs on either side of the river are from sixty to a hundred feet in height, and the stream flows from side to side of the valley, leaving flat bottom-lands within its curves of varied extent. Where we struck the river one of these rich meadows was found, containing some two hundred acres, skirted on the opposite side by the river, fringed with cottonwood trees, whose rich green color was a gratifying contrast to the brown and withered vegetation of the region we had just left. Encircling this flat on the side next to the bluff was a growth of timber bordering a slough, across which we had to construct a roadway, with considerable labor, for the teams.

AUGUST 13.—No march was made to-day. The surveying party ascended the river seven miles, to a point where the survey of last year terminated. Having found the last stake [from the 1871 survey], the engineers commenced work, and carried their line upon the broad terrace next above the river bottom, and continued it to a point nearly opposite this camp. Our camp is exceedingly pleasant and quite picturesque. It is entirely surrounded by trees, among which is a rank growth of willow and rosebushes. The blossoms of the latter have fallen some time since, leaving almost as beautiful a display of bright red berries in their stead.

Close to camp rushes the river in a perfect torrent, washing away the soft alluvial bottom with great rapidity, while evidence of its former power is seen in the huge trees that lie stranded on the small islands in its midst. The stream is here about two hundred yards wide. The opposite shore is a bold, rocky bluff, perhaps eighty feet in height, worn by the elements into various shapes, here and there resembling battle-castles of the feudal ages.

Newspaper Accounts

Helena Daily Herald, July 10

OUR EASTERN MONTANA LETTER
TO THE EDITOR OF THE HERALD

BOZEMAN, JULY 7, 1872—On the 20th inst. Col. Roberts[51] will start down the Yellowstone with an engineering party. They are to have an escort of the four companies of cavalry at Ellis and four of infantry from Shaw, all under the command of Col. Baker. Their intention is to follow the Yellowstone down to the mouth of Powder River, where they will connect with the surveyors from the East. From this point they will turn back, cross the country to the Big Bend of the Musselshell and follow this river to its source. I believe it will only be a preliminary inspection of this route as the time would be too short for an actual survey. This expedition shows, whatever may be its result in regard to the choice of route, that the view I took last winter in my article in your paper was correct, that the company would not select its route without making further examinations of the country. We shall gain much interesting and useful information about a portion of our Territory which is yet almost entirely unknown, and I hope it will lead to the adoption of a final route this winter, so that we may know what to expect.

In spite of the snow storm on the 1st, fresh strawberries are the order of the day.

Yours truly, PK

Little was to have more impact on the 1872 Western Yellowstone Survey than the height of the Yellowstone River that summer. While most of the below article concerned the 1872 Yellowstone Park Expedition under Ferdinand V. Haydon, some of it is key to understanding the Northern Pacific's subsequent troubles.

---

[51]The writer likely confused Milnor Roberts with his son Thomas, the latter being then in Montana.

Helena Daily Herald, July 26

YELLOWSTONE CORRESPONDENCE
TO THE EDITOR OF THE HERALD

Crow Agency,[52] July 23, 1872—The flood of the Yellowstone—fully one month later than last season—has now fairly commenced subsiding. Why, that little pestiferous institution of the genus culex—the mosquito—subsided also.

Plenty of snow in the mountains yet, causing rapid condensation of the mists arising from the heated soil, and resulting in frequent thunder showers; consequently we are having a wet season, with abundant grass on the bottom-lands, benches and foot-hills, looking as green and luxuriant as in May. This plentiful growth affords encouragement to stock men.

[Signed] Yellowstone

The following contains one of the few mentions of the western survey's departure. The lack of information reflects Montana's small 1872 population (perhaps 23,000) and only two regular newspapers that followed the story. Although not stated in these articles, there was significant eastern Montana opposition to the survey. And why was this, considering the state's desire for a railroad? With small but continuing Indian (mostly Sioux) raids into the Gallatin valley, Fort Ellis was left with fewer than thirty infantry officers and men, many of them physically unable to join the surveying expedition.

Bozeman Avant Courier, August 1

GONE

The North Pacific Railroad surveyors, under Col. J. A. Haydon, left on Sunday morning for the Yellowstone to complete the Survey of that country and run the line on the iron horse that is soon to come screaming to Bozeman. The military escort, consisting of nearly all troops stationed in Fort Ellis, is under the command of Col. Eugene Baker.

---

[52]The Crow Indian Agency was located on the eastern/southern side of the Yellowstone, just east of the Bozeman Pass and present-day Livingston, Montana.

The Northern Pacific in Trouble

Far too many studies of military events—which is what the western surveys proved to be—omit virtually all mention of business and financial matters. Certainly the Northern Pacific's two Yellowstone surveys were not conducted in an economic vacuum. While financial matters were of zero concern when surveyors and their army escorts heard war whoops, 140 years ago the Northern Pacific was in a state of collapse due to vacillating and corrupt management, the insufficient sale of bonds and land, and the increasingly limited resources of Jay Cooke & Company.

The following section begins with Northern Pacific president Gregory Smith's forced letter of resignation. Despite all his maneuvering to get Smith out, Cooke then had no replacement. Smith, meanwhile, changed his mind, and since his resignation had not become public, he continued to operate the company for another four months!

However, by far the most interesting document is Jay Cooke being given the Northern Pacific's business facts-of-life by his hard-nosed protégé, Harris Fahnestock, the financier's exceptionally bright number two man.

* * *

Jay Cooke's firing of Northern Pacific president J. Gregory Smith was no easy task. Smith had numerous allies on the board of directors and had made considerable money for them (and Cooke) through land speculation. The real issue was that Cooke—with good reason—no longer trusted Smith, and he used the issue of Smith's devoting his full time to the Northern Pacific or resign. Smith's resignation follows.

J. Gregory Smith to Frederick Billings, May 15

Hon. Frederick Billings

Dear Sir,

I am this day in receipt of a letter from Mr. Cooke urging upon me the necessity for giving my whole time to the affairs of the [Northern Pacific] and requesting me to resign my connection with the [Vermont Central].

I fully recognize the fact that it is essential for the best interests of the Northern Pacific that it should have at its head some gentlemen with less burden upon him than I have and who can give his undivided time and energies to this great enterprise. I have only been deterred hitherto from resigning my position in connection with this road in the fear that such a step might become a disturbing element in the then condition of its affairs.

Now, however, that its prospects are more hopeful and it seems to rest upon an established basis, there exists no longer any reason why I cannot with safety to all interests accomplish for myself what I have long desired to secure and what my family and friends have with great seriousness urged upon me—a respite from the excessive labors and responsibilities which have hitherto devolved upon me.

I will thank you as Chairman of the Committee on Salaries to do me the favor to call your Committee together at an early date to determine that question that I may be in condition to close my relations fully with the company.

I shall esteem it a personal favor if you will comply with my wishes in this regard at an early date.

[Signed by Smith]

Source: "Secretary's Department: Unregistered Letters Received . . . 1864–1876," microfilm M-459, Minnesota Historic Society, St. Paul.

Harris Charles Fahnestock (1835–1914) had been a bank "clerk" in Harrisburg, Pennsylvania, until coming to the attention of Jay Cooke in 1861. In that era, however, a clerk's responsibilities were far-ranging and important as numerous regional banks and even some companies printed their own paper money, an open invitation to counterfeiters. Additionally, gold and silver coinage could be rubbed down or recast with cheaper metals. Thus it was up to the head clerk to quickly determine the value (if any) of the cash being exchanged in a particular transaction. Fahnestock immediately impressed Cooke, who first "borrowed" him for a sale of Pennsylvania war bonds, then placed Fahnestock in practical charge of his newly opened Washington office—the official nominally running this bank being Cooke's brother, Henry. After the war's conclusion, Cooke had Fahnestock independently manage his vitally important New York branch. Not held back by his ice cold personality, Fahnestock had a mind of his own and by 1872 knew he was far and away the company's best banker. At this same time, now thoroughly disillusioned by the Northern Pacific's troubles and Cooke's increasingly autocratic management style, he respectfully but candidly laid out the unpleasant financial truth to Cooke in this letter.

Harris Fahnestock to Jay Cooke, June 8

Dear Jay:

Whilst very reluctant to disturb the pleasure of your vacation by the discussion of business, I feel it important to bring to your attention the pressing necessity of certain immediate reforms in the organization of the N.P.R.R. and changes in our relation to the company which I consider to be <u>vital</u> to both them and us—matters of policy to which you can better now give consideration than when in Phil. or N.Y. and constantly full of engagements.

I beg that you will not regard any of my views & criticisms as personal reflections, but they shall be received as the earnest expression of my convictions after careful study of the situation.

I do not hesitate to say that the present actual condition of the N.P., if it were understood by the public, would be fatal to the negotiation of its securities. The fact that the construction of the road, the details of its management and the disposition of its lands are in the hands of men who are so fully occupied by other interests that the entire enterprise is practically without a head, excepting yourself, who are known to be absorbed in the provision of its finance, would be enough to disturb confidence; but the added facts that there is no provision whatever for the July interest upon the Company's bonds and that the payment of the coupons depends solely upon our ability and willingness to protect them, and that the prosecution of the work and payment of the maturing contracts for labor & materials are all dependent upon financial provision yet to be made, would, if known, be sufficient to destroy confidence entirely.

No enterprise of such magnitude has ever before been so entirely dependent on one house, or rather on one man. . . . But when, wisely or unwisely, our house has been so closely identified with the N.P. it behooves us to see that our statements, promises and guarantees are justified and fulfilled to the letter so that in the event of misfortune we cannot be held even morally responsible to the innocent bondholders; and, on the other hand, that we shall not be personally ruined. . . .

Experience has shown it to be impossible for you to personally manage the details of the construction of the road, and that Gov. Smith's management has been inefficient, and the determination to substitute Gen. Cass is a most important step in the right direction. Cass's reputation as a railroad man will to a great extent stand between us and the public, but it is indisputable that the reform shall be thorough in every department, and that the organization shall be complete and self-dependent: that the most rigorous economy shall be introduced and that the road shall be placed in the hands of men who can and will carry it on independently of us. . . .

[W]e, and especially you, are the N.P.R.R., and . . . the bonds have been sold to a class of investors who have been influenced by your personal recommendation.

We have assured them of the intelligence, vigor and economy of the management, [but] we know that it has been inefficient, distracted by other engagements and extravagant to the last degree. We have assured them that the lands are unparalleled in climate, soil, timber and [m]inerals. . . . [Moreover,] we know that a large proportion of the lands from Lake Superior to the Mississippi are practically valueless either for cultivation or for lumbering and the residue are less valuable than the public have been led to believe.

From the Mississippi to the Red River our statements are sustained by the quality of the lands. From the Red River [and beyond] . . . we do know however that a large portion of it is worthless and the reports of our engineers . . . are, to say the least, anything but flattering.

Our publications have given promise of rapid marketing to the lands for the security of the bondholders. Do you know that after all has been said and done in the organization of the Land Dept. the aggregate cash results [is now only] $388.76. Our publications have enlarged upon the comprehensive and perfect system of immigration which is to fill up the lands as the road progresses, thus making their value and their market. I have yet to learn that any definite steps whatever have been taken in this direction or . . . accomplished any results worth mentioning.

In short, although the success of the enterprise is almost entirely dependent upon the development and settlement of the lands, these two most important features [under Smith] . . . has been so neglected that practically nothing has been accomplished. . . . When you refer to the experience of the Lake Superior & Miss. R.R. both in respect to land sales and immigrants you can form some estimates of the Herculean work [that needs] to be accomplished in the same direction by the N.P.R.R.

Our publications (even down to the last pamphlet) have too extravagantly painted the whole enterprise and their effect on the business men, capitalists & people whose judgment governs their investments has been prejudicial to the securities. Too much dependence has been placed upon the names of the promoters, and too little weight given to the more practical considerations which must govern the public estimate of securities offered for investment. . . .

Questions will presently arise about the traffic of the road. All other roads with any pretensions to public confidence count upon some traffic returns even during the course of construction. We have practically none and can expect none to speak of before reaching the Missouri River. Even if we were through today to the Pacific we must await the creation of a commerce which we cannot expect to suddenly divert from a great city like San Francisco.

The Union Pacific line, with all its advantages of established trade, has several times been in deep water and has only been rescued by contributions of its large wealthy share-holders, and today if forced to pay the five millions interest it owes the U.S. it would be bankrupt. [While] we are building a road without money aid from the U.S. . . .

Without any disposition to underrate your enthusiasm which has forwarded the enterprise and has accomplished most of what has been done (and indeed without which it could not have been done) we must openly look in the face [at] all errors of the past and see what is now best for the road and ourselves. Enthusiasm alone will not be sufficient; it must be combined with the soundest judgment.

I have showed you why the general public will not buy the bonds . . . because everybody knows that their value depends on one man's ability to make them good. . . . All other important roads have the aggregated responsibility of many good men, reasonable certainty of completion within a moderate time, objective points possessing elements of immediate or early revenue, and prices bearing some relation to the risk.

We ignore most of these considerations and the enterprise is new, with all our delicate relations to the confiding public, our peculiar relations to the pool subscribers who rely upon at least discreet management upon our part, and to the contractors for labor and materials, completely at the mercy of any contingency[53] which may arise to interrupt sales of bonds, stop the work, and leave us to provide means to save the company from default—entirely alone or by humiliating appeal to the pool subscribers.

I have thus hastily thrown together the prominent defects of our present situation. Radical and immediate changes are necessary to save the company from ingloriously breaking down within the next year and involving us in discredit if not in ruin. We must be so divorced from all but the finances that we shall not be responsible for the mistakes of the Company, and we shall not do justice to the public or to the bondholders until the progress of the work is gauged strictly by the money absolutely provided and we can always have in hand a fund sufficient to guarantee the interest for at least a year ahead. . . .

You have observed the great falling off in sales, I do not despair of a revival. . . . But I am alarmed with good reason by what I see and, including the large advances we shall have to make for interest; the Company will be in our debt on 1st July at least two millions—probably 2-1/2. . . .

With [some 250 miles] of line from Duluth we shall next spring reach the Missouri with a bonded debt of upwards of 20 millions and practically no receipts from any other source.

The purpose of this letter is to impress upon you the precarious present situation and to set you to thinking about the best method of accomplishing what I regard as indispensably necessary for ourselves. The road can be finished to the Missouri, but beyond that point I think it must depend upon an entire rearrangement.

[Closing words unclear, signed by Fahnestock]

Source: Harris C. Fahnestock, Miscellaneous Letters to Jay Cooke, October 1868–April 1874, F-223, box 3, folder 2, Baker Library, Harvard Business School, Cambridge, Mass. All emphasis in original. Given the letter's length, numerous substantive cuts have been made, hopefully without reducing the urgency of its tone.

[53]Not stated, of course, but in the back of everyone's minds was open Indian warfare, an issue which had not formally developed when this was written.

Cooke to Smith, August 6

Dear Gov. Smith:

Just telegraphed you that I would write tonight, and that I would like to have an answer tomorrow by telegraph.

I observe by an abstract of the proceedings, which the Secretary[54] has sent me as usual, that many things are left unattended to. I want to talk all these over with you thoroughly. When will you be in New York to remain several days[?] I will go over next week, Monday, or earlier. Would prefer Friday of this week if you will be on hand. Things are now in somewhat of a chaotic shape in regard to the committees proposed, the Lake Superior wheat business, St. Paul and Pacific [Railroad], etc. I want to have a good talk with you over all these matters and especially over the finances.

I find that the drafts are swelling up the overdraft to nearly a million and a quarter. We cannot stand this of course, as we are under no obligations to advance beyond a half million. The visit to the Pacific coast will postpone of course for a long time any meeting at this end, most probably. Before the Directors go so far away they must look squarely in the face the whole financial question, and they must give time enough to it to decide upon measures to save money. We are willing to do our share but cannot do everything in the face of the fearful amount required if the work is to be pushed forward as rapidly as [Milnor] Roberts telegraphs; viz., 2-1/2 miles a day.

Will you please call a meeting of the Board for next week Tuesday or Wednesday? Do not fail to do this, dear Gov., for it is an absolute necessity, I assure you, in view of the financial situation. It will give time before the Directors leave for the Pacific to make any necessary arrangements, and to work up the other committees proposed.

I enclose a letter from Frank Clark in regard to the wheat business. There ought to be an effort made at once to arrange with the Chicago & Milwaukee roads. This whole matter ought to be put in the hands of a working committee who will have time to follow it up to a final conclusion, including such negotiations as shall insure a fair share of

---

[54]Sam Wilkeson (1817–89) had been a close wartime associate of Cooke's and was secretary of the railroad for more than twenty years, his duties including all advertising, public relations, and related marketing functions. With Wilkeson, the usually inhibited Cooke began his letters "Dear Uncle Sam" or "Samuel."

business over the Lake Superior Road. I believe this can be done but only in the way I propose.

Yours Truly, etc.

Cooke to Smith, August 9

Dear Gov. Smith:

Yours of the 8th received with enclosures. Please let me hear from you tomorrow, if you are in New York, especially as to the date of the meeting which I requested in my last letter sent [to] you [in] Boston. Telegraph tomorrow, if you can, the date you fix upon. Please let it be as early as possible so as to give as much time to the meeting as parties can spare. There is much to be talked over and arranged financially as well as in regard to [the] St. Paul & Pacific which is a vital matter staring us in the face.[55]

The matter of the interest upon those bonds due the 1st May I trust will be settled according to the plain understanding. [Jay Cooke &] Company has no resources, and did not provide for that. The bonds were maturing and remained maturing until the close of the bank houses on the 1st. As the contract was signed earlier than that in the day, if it is necessary to stretch the understanding at all, that would fix it all right, but the plain understanding was that the interest coming due was to be paid by the lessees to be reimbursed out of the earnings and the first proceeds of lands [sales]. There will occasionally be some back coupons coming in. These ought to be paid in the same way, as the whole power and finances are in the hands of the lessees.

Yours Truly, etc.

Cooke to Smith, August 12

Dear Gov. Smith:

Yours of the 10th received. I telegraphed you asking if you could not be in New York a day or two before the meeting of the 20th so that I

---

[55]Cooke had taken control of the St. Paul & Pacific, but under the direction of his brother-in-law, William Moorhead, a bad financial position had been made significantly worse. By this point Cooke was unable to pay the Dutch investors who had originally financed the road, and he was about to lose control of it. For full details, see Augustus J. Veenendall's *The Saint Paul & Pacific Railroad.*

might see you and cut and dry a good many things that will save time when the meeting comes together. *The finances must be put in shape or you cannot go to the Pacific Coast* [emphasis added].

I notice what you say about the wheat question. I have had a talk with Mr. Potts[56] and expect to see him again today. He seems to be willing to make some arrangements in which the steam-boating will participate, also by which neither the steamboats nor the railroads will run any risk in speculation, and will merely concede a certain portion of the freight by way of encouragement. He thinks parties can be found who will make the purchases by some such concession, but his best suggestion is that the matter be taken hold of at once, and the opposition railroads and buyers be conciliated by a proposition of this kind, to let them have the whole field to themselves, they agreeing to send by way of Duluth, say, one-third of all the wheat exported from the State on terms agreed beforehand. This would make an amicable settlement and would be very profitable as they could put their own price upon the wheat in Minnesota without opposition. If they put the price too low, other buyers would come in and the farmer would be protected. But at a fair price they would get hold of it.

Since writing you Mr. Potts has been here and it is understood that they will cooperate. They will send Mr [——] of Buffalo to Milwaukee to meet Mr. Mead[57] and the representatives of the Milwaukee & Chicago roads for the final settlement of this question. They will cooperate on either of three plans:

1. Giving the wheat [——] of the Milwaukee & Chicago full control of all purchases on the basis of their deciding the carrying trade, a portion of which must go through Duluth on fair terms.
2. Any arrangement to purchase in fixed terms and divide the purchases, a portion to go to Duluth.
3. A regular fight until they get sick of it.

One of the largest dealers in the city, Peter Wright & Sons,[58] who can control lots of money, Mr. Potts thinks will readily go in under either

---

[56]Benjamin F. Potts (1837–87), from Ohio, was appointed territorial governor of Montana in 1870 and served thirteen years. He began the war as a lieutenant and finished as a brevet major general, impressing all with his determined leadership. Potts was a Cooke stalwart, supporting the Northern Pacific at every juncture.

[57]G. W. Mead was the Northern Pacific's general manager of operations in Dakota and Minnesota.

[58]Peter Wright & Sons was a Philadelphia shipping firm.

of the two last plans, if the road, [grain] elevator and the steamboat companies will make up a reasonable concession. Should I telegraph you tomorrow at the request of Mr. Moorhead, Mr. Potts, Mr. Wright and others to have Mr. Mead go to Milwaukee at once you will know what it means. Expect Mr. Newcomb here by tomorrow. Shall then see if all these parties again [words missing?] and shall telegraph you fully.

Yours Truly, etc.

Source: NPRR *Secretary Letters, Unregistered Letters Received, 1864–76, Northern Pacific Railroad Files, Minnesota Historic Society, St. Paul.*

PART II

Fighting on the Yellowstone

August–September 1872

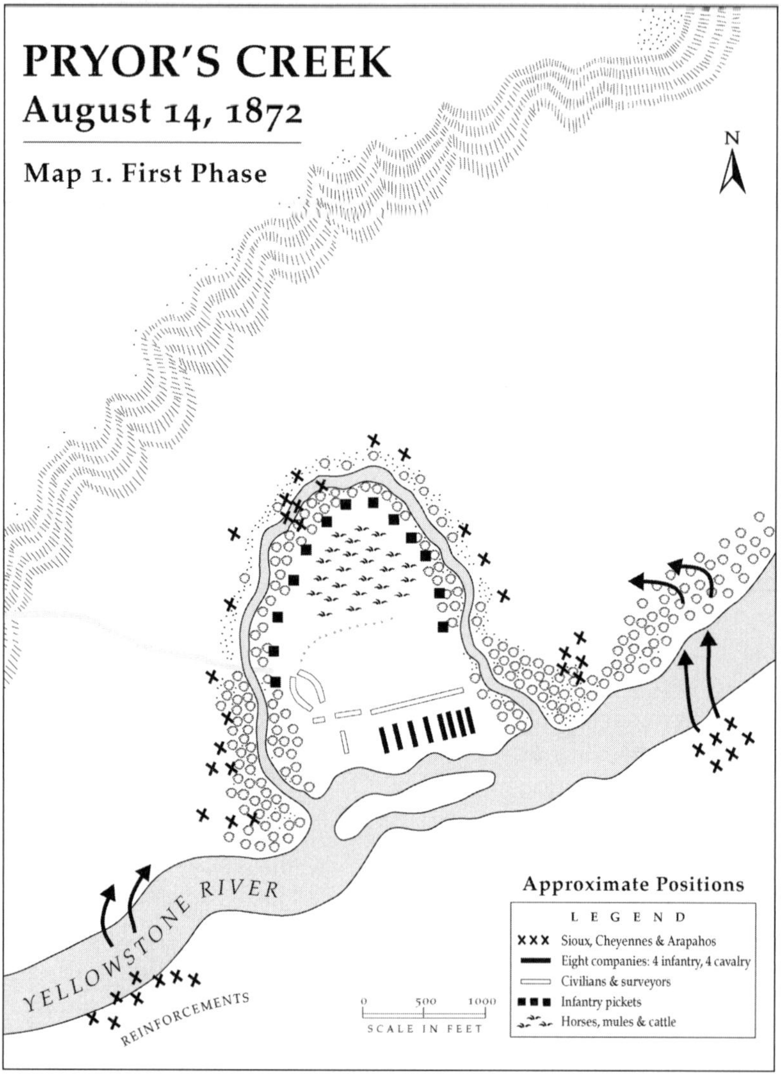

Map by Vicki Trego Hill, adapted from an 1872 map likely drawn by Maj. John W. Barlow, courtesy of the National Archives and Records Administration, Washington, D.C. *Copyright © 2015 The University of Oklahoma Press.*

Progress by Stanley and Rosser across western Dakota was comparatively rapid. Although summer rains had flooded streams and softened the prairie, less than two days' travel was lost, an insignificant disruption. The 1871 route was followed until ten miles east of the Heart and Green River junction at Gladstone, North Dakota. Here they cut southwest, reaching the Little Missouri fifteen miles above Marmath, North Dakota, on August 9.[1]

Rain made the river temporarily impassible, but everyone enjoyed the break, many finding *petrifications* (dinosaur or plant fossils). On August 11 Rosser, who had crossed rivers under fire in his Confederate days, became incensed when Linsley began second-guessing him. Otherwise without incident, the column crossed the river, then laboriously cut a path up and alongside the mosquito-infested steep hill just to the west.

Here there is a remarkable change in geography, for the ground undergoes a startling metamorphosis to arid, almost desertlike conditions with few trees and limited shrubbery, and any water—when found—is as often as not alkaline. Despite this, the firm ground permitted rapid work, the column reaching O'Fallon Creek on August 15.

* * *

If Stanley and Rosser had limited problems, the same could not be said for the western survey. Their plan was to cross to the Yellowstone's east/southern bank just below Livingston, Montana, a location easily waded. However, late spring rains and a rare July snowstorm saw the Yellowstone higher than anticipated.

Next came Baker's fateful decision that doomed the survey. Following a likely scary crossing on a permanent but light rope ferry, rather than wait for the river to fall or build a larger ferry, Baker decided to have his column march along the difficult northern bank, expecting the river to be lower opposite Big Timber, Montana. To everyone's horror, this proved impossible. The result forced the column to cut north across rugged country toward the Musselshell River. Late August 12, after a backbreaking week, the column returned to the Yellowstone, just east of Billings where the last 1871 stakes had been placed.

Baker's column was now not only on the wrong side of the Yellowstone but over a week and one hundred miles behind schedule. Undoubtedly

[1]Today, at this same location, a dirt road reaches the Little Missouri at a low concrete bridge.

under increasing tension, he had begun drinking. Once at the Yellowstone the exhausted column collapsed after finding an idyllic campsite with high trees and tall grass. Yet, despite numerous Indian signs, no one exhibited concern. About 3 A.M. August 14, as twenty-six men lazily (many undoubtedly sleeping) guarded the camp's perimeter and a half-dozen officers played poker, hundreds of Sioux, Cheyenne, and Arapaho began crawling toward them.

Along the Powder River near the Montana-Wyoming line, a few thousand warriors, many with their families, had begun gathering in July.[2] One person missing was the Hunkpapa war chief Gall. Earlier he and a few dozen warriors left Sitting Bull and rode toward Fort Rice. When they first spotted Stanley is unknown, but Gall undoubtedly sent word back to Sitting Bull concerning the number of soldiers, the fact that they were infantry, and the type of artillery.

Here, the lack of primary sources requires supposition and speculation. It is believed there were two thousand or more lodges on the Powder River, and one assumes that the reason for their gathering—although there is admittedly no *evidence*—was to block the Northern Pacific.[3] Sitting Bull, while leading those opposed to the railroad, nevertheless led by consensus. When Gall's scouts returned, a plan of action was formulated. Their encampment was (coincidentally?) located roughly equidistant from where the two surveys entered the Yellowstone Valley. What the Indians had to decide was whether to attack one or both columns, attack the Crow, or do nothing.

While Baker's trek toward the Musselshell temporarily hid him from the Indians, Sitting Bull knew the column would go to 1871's termination point near Billings. Baker had no artillery, and his column also contained hundreds of cavalry horses worth stealing, a huge incentive. Also, the Indians had to be aware that Baker, whom they hated, was leading the column, another motivator. To what degree, if any, they knew of Baker's overconfidence is unknown.

Ultimately, after debate, the decision was made to attack Baker and leave Gall to harass Stanley. Stanley, without cavalry, posed no *offensive*

[2] Utley, *The Lance and the Shield,* 107.

[3] The predetermined location wouldn't have been made to hunt buffalo, and its exceptional size would not have been achieved simply to make an attack on the Crow.

threat. Additionally, attacking artillery represented the most frightening option for Indians, and they had no wish to do so again. The final reason is that the Indians were sophisticated enough to analyze Stanley and Baker's capabilities. Thus, quite sensibly, the higher-risk, lower-reward option was rejected.

Later the Indians concocted a story in order to receive winter rations; Indian agents were told that their goal had been to attack the Crow agency near Livingston, but finding themselves near Baker's camp, they decided to steal his horses. Contradicting this is that when Sitting Bull and perhaps a thousand warriors broke camp in August, they rode directly for Billings, where they found Baker as expected. However, if they had ridden a more southern route to keep hidden from the Crows, they would have missed Baker entirely.

When Sitting Bull's scouts spotted Baker on August 12, his warriors were south of the Yellowstone, likely near Pryor's Creek, behind the hills between the creek and the river, hidden from Baker. Despite numerous Indian signs, Baker sent no scouts across the river on August 13. If anything, Sitting Bull was surprised to learn that Baker's column was on the north side of the Yellowstone. Certainly in the week that followed, many of the Indian actions and taunts are inexplicable other than in the context of this belief.

Late on August 13 the Indians gathered to discuss their next step, but at the last minute, older leaders had second thoughts about swimming the Yellowstone. However, as discussions droned on after dark, younger warriors gathered their weapons, slipped out of camp, and swam the Yellowstone without waiting for permission. Sitting Bull, facing this de facto exodus, now had his own to choice to make; lead, follow, or get out of the way. He soon swam the river, but in the morning's wee hours there was no way to execute a coordinated plan. Predictably, the Indians were spotted during their piecemeal arrival, and the surprise attack fizzled.

* * *

To the east, surveying continued without incident as the column neared the Yellowstone. Stanley had no way of knowing about Baker's August 14 battle, but news traveled quickly: army officers called it the "Indian grapevine," akin to World War II's "jungle telegraph." Thus Gall, with his few dozen Hunkpapa, was soon aware of the Baker fight. At dawn, August 16, they galloped by the surveyors' side of Stanley's camp firing

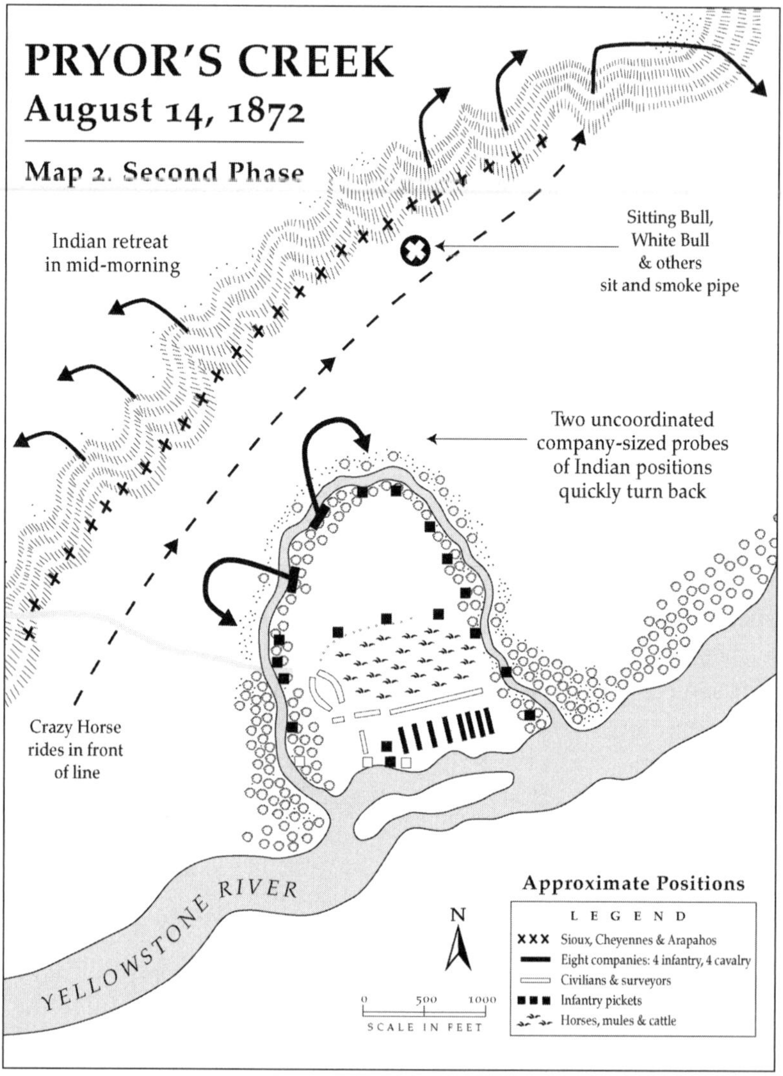

Map by Vicki Trego Hill, adapted from an 1872 map likely drawn by Maj. John W. Barlow, courtesy of the National Archives and Records Administration, Washington, D.C. *Copyright © 2015 The University of Oklahoma Press.*

shots and shooting arrows. While hitting nothing, they let their presence be known and demonstrated their bravery. Three army mules were hit, one apparently shot through the head by the surveyors. Stanley, not amused, had the surveyors sleep next to the wagons, put a company of infantry around them, and each night confiscated their rifles—although, one can rest assured, not Rosser's.

The Yellowstone was reached on August 17. The next day the surveyors and half the soldiers marched to the Powder River, thereby accomplishing their basic engineering goal. Thirsty army officers and surveyors broke out two cases of champagne, quickly drinking it in 100-degree heat. This thoughtful planning went back to 1871 when, on reaching the Yellowstone, Rosser and army officers agreed that they had forgotten to bring champagne—a mistake not repeated. By noon the surveyors finished working and placed a wooden "zero-mile" stake for the western surveyors whom they assumed would arrive shortly.

Earlier, surveyor Henry Davis had slipped away and was hunting alone for agates when he was seen and chased by Gall. Davis, as he relates in his diary, was unable to tell the difference between the Rees, riding to his rescue, and the Hunkpapa. The chase quickly led to a series of clashes between Gall's warriors and the Rees, which Rosser clinically describes in "The Death of Adair."

That same afternoon, Stanley, taking his life in his hands, rode into the shallow Powder River to meet Gall. The discussion led nowhere, but Stanley first heard the news of Baker's battle, which Gall regarded as a great victory. Equally frightening, Stanley learned that Baker was on the "wrong" side of the Yellowstone and that Sitting Bull would join Gall with massive reinforcements. The next evening Stanley cracked. In his diary Rosser wrote, "Commanding Officer shows signs of going on a spree."

Once the surveyors' work was completed, Stanley, correctly believing Sitting Bull would attack, began his return. Most frightening for Stanley, the only practical route was to retrace his steps through the half-mile-wide O'Fallon Valley, a death-trap if Sitting Bull had the force Stanley anticipated.[4] But when Sitting Bull did appear, he had at most two hundred warriors and little ammunition.

---

[4]The valley's walls rise 150 to 175 feet above O'Fallon Creek. Standing on the valley's floor and looking up, Stanley had good reason to be fearful.

In the skirmishing that followed, at least one Gatling gun was fired against mounted Sioux who charged Stanley's rear. Hollywood adventure movies notwithstanding (e.g., *Gunga Din*), the Sioux, galloping over broken ground, were moving too fast for a Gatling gun to be effectively traversed. Ultimately not a single Indian was hit.

* * *

Meanwhile, upriver, by the afternoon of August 14 the western column was again under way. While the Indians had been defeated by army standards,[5] Baker (too drunk to command), his officers, and chief surveyor John Haydon were badly shaken. That night a report was sent to Fort Ellis and Col. John Gibbon, while engineer John Haydon sent a frantic note to Northern Pacific officials. Communiqués were sent not just to army personnel but also to railroad officials, the *Bozeman Avant Courier*, and the *Helena Herald*. All messages were given to a small detachment which slipped out of camp that night and reached Fort Ellis–Bozeman during the night of August 17–18.

What happened next proved to be controversial but ultimately disastrous for the Northern Pacific. After marching three miles on August 14 and two and a half the next day, Baker refused to move further. On all other Yellowstone surveys, the army daily paralleled the surveyors to protect and minimize their walking. Now Baker forced Haydon to go further alone each day, with only one cavalry company (perhaps forty-five men) escorting them. On August 19, in terrible heat and humidity, Haydon and the surveyors marched some six miles before *resuming* the previous day's surveying. After surveying two and a half miles, frightened by Indians who were in constant sight on the Yellowstone's eastern shore, Haydon gave up. Returning to camp exhausted, he met with the rested Baker and signed a brief letter, dictated or given to him by Baker, which requested that Baker turn back. This writer suggests that all Baker's actions were calculated, forcing Haydon to request that the expedition return.

On September 11, the recriminations became public, with the War Department blaming Haydon and the Northern Pacific. Neither the railroad nor Cooke dared to engage in a public dispute with the army, especially with Grant seeking reelection.

---

[5]No contemporary source discusses the Indians' frequent acts of bravery, notably those of Sitting Bull and Crazy Horse. Only Lt. Edward J. McClernand mentions this in *On Time for Disaster* (29–30), which he wrote fifty years later.

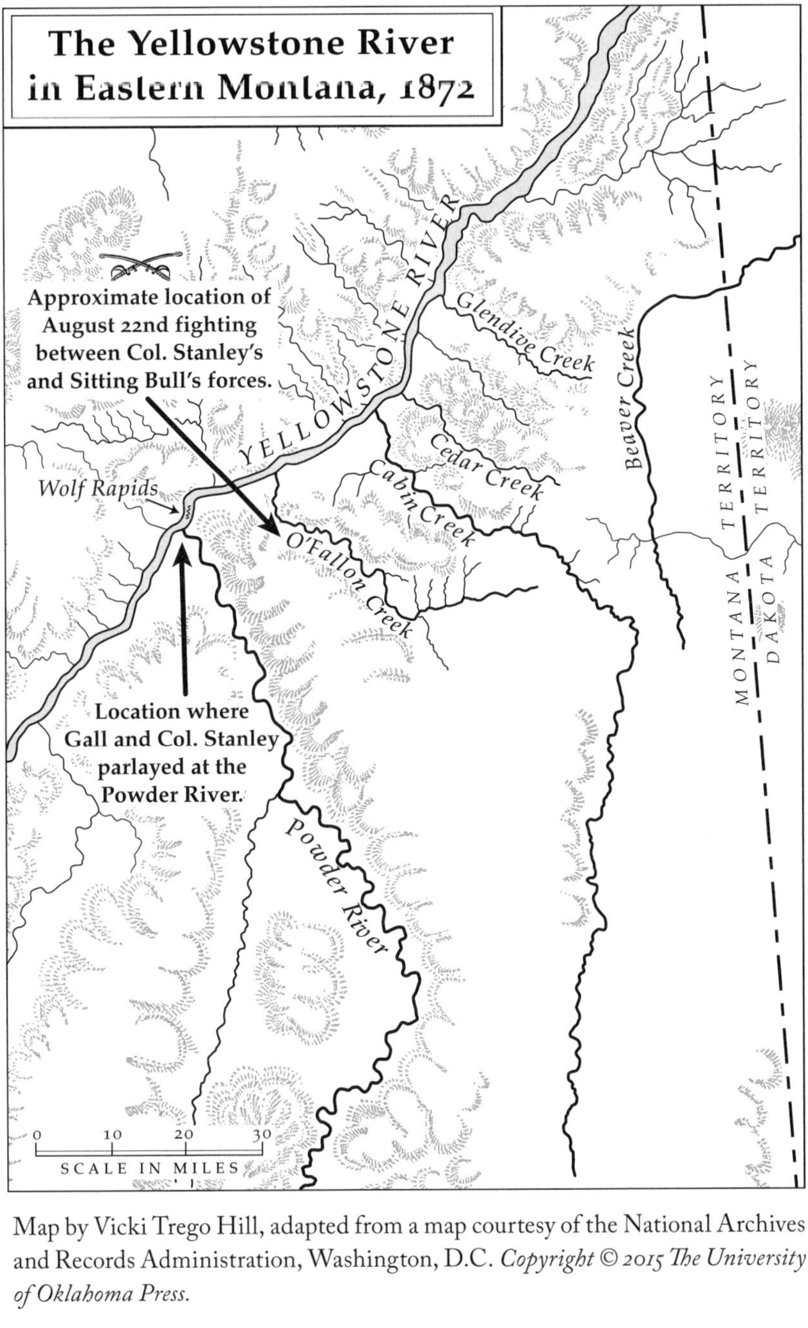

Map by Vicki Trego Hill, adapted from a map courtesy of the National Archives and Records Administration, Washington, D.C. *Copyright © 2015 The University of Oklahoma Press.*

Eastern Survey

The Death of Adair (excerpt)

The following June [1872], having secured a suitable escort by command of Gen. Winfield Scott Hancock, a larger party of engineers was assembled at Fort Rice early in July with rations for a three-month campaign. Gen. D. S. Stanley, a graduate of West Point and a distinguished Major General in the "war between the states," was found there with orders to organize and command the escort. Stanley, in command of the military district of Dakota, was thoroughly acquainted with the Indian character and mode of warfare, and I felt exceedingly fortunate in having so able an officer in command. The escort embraced detachments from the Sixth, Eighth, Seventeenth, and Twenty-second regiments of infantry, a battery of [three] Gatling guns, and a hundred friendly Sioux Indian scouts.[6]

This expedition crossed the Little Missouri River at the great bend and reached the Yellowstone at the mouth of O'Fallon's Creek. We had passed over a different country from that traversed the fall before. Bearing further south we avoided the Valley of the Heart, and passed above the sources of Cedar and Cabin Creeks, making a march of about 300 miles. In this distance we saw great quantities of game, lazily feeding on extensive and beautiful prairies which no white man had ever crossed before. But we saw no Indians, indeed saw no signs of any and in amazement we asked, "Lo, where is the poor Indian?" In answer to these

[6]Here is a good example of a person's memory playing tricks on the writer; the actual number of Indian scouts was no more than thirty, few of them Sioux and none of them Lakota.

questions, General Stanley, who knew the Indian character well, would reply, "Doubtless he is not far off, and is only watching his opportunity to surprise us." But not withstanding his warnings, we grew careless and hunting parties recklessly wandered far from the marching column.

The day before we reached the Yellowstone, Lieut. Ketchum[7] who was adjutant of the command, Captain Kellogg,[8] who was serving on General Stanley's staff, our guide Clement, myself and the mounted scouts rode ahead as far as the Yellowstone, then returned and waited on O'Fallon's Creek for the column to come up. There were a few cottonwoods trees along the creek, the day was warm, and all of us hobbled our horses and carelessly threw ourselves in the shadows of the trees and went to sleep. The sun moving the shadow from me I awoke and when I saw everyone around me asleep I became frightened and called "to horse" and I can never think of our recklessness without a shudder.

The command reached us just before sunset, and went into bivouac where we had been taking our careless siesta. Our camp was upon low ground, and the bluffs were on each side of us and not far off. The sky was clear and starlit, but there was no moon. Sentinels were posted as usual, and an officer put in charge. The night was passing peacefully away. A faint glimmer of light just touching the eastern horizon announced the approach of another day, but the usual reveille did not proclaim it. A volley from the hills and the wild demoniacal shout of attacking savages called us from our slumbers. The howling of hungry wolves at night in the wilderness is indeed dismal, but the wild "hi-I, hi-I" of savage men in war paint and hideous disguises breaking in upon your slumber in the death stillness of the night, for a moment paralyzes you, then braces every nerve in your body for defense. Gen. Stanley was prompt in rallying his men and cool in commanding them, and soon drove the savages off. After this little flurry, there was no more sleeping, every man walked about with his musket, and every officer and citizen put fresh charges in their rifles and pistols, and assured themselves that their action was perfect.

---

[7]Hiram H. Ketchum (1843–98) was Stanley's adjutant (duties similar to today's administrative assistant) on both the 1872 and '73 surveys. He enlisted as a private in 1861 and retired as a major the year of his death.

[8]Capt. Sanford C. Kellogg (1842–1904) was an in-law of Gen. George H. Thomas and Sheridan's personal observer.

Daylight soon came and Gen. Stanley moved a line of troops out and occupied the hills, but the Indians having failed to surprise us, had withdrawn to await another opportunity. We marched down O'Fallon's Creek, and encamped on a commanding plateau overlooking the beautiful Yellowstone. A survey had been ordered from the Upper Yellowstone, and it was expected that we would meet them at the mouth of Powder River which our guide informed us was about six miles further up the River. Leaving the camp unbroken and well guarded the next morning, Gen. Stanley took a battalion [or half his] infantry, a section of artillery and the mounted scouts and marched to the mouth of the Powder River. The stream was bold, turbid and flushed by the recent rains in the Black Hills, and was too formidable to be crossed by a command unprovided with a pontoon team.[9] It was not absolutely necessary for us to cross as the country was all prairie, and we could see that our friends [from the western survey] were not there nor could we see any signs of them. The grade of the Yellowstone River, and all of its tributaries, are steep. They are shallow and rapid and flow through a narrow valley which they fully occupy during freshens, but at low water the channel runs from one side of the valley to the other, colliding against the hills which they cut into perpendicular bluffs. The spaces lying in the bends are usually covered with a heavy growth of cottonwood trees, but nowhere else in the region are trees to be found.

There are here and there a few scraggly pines and red cedar up in the cliffs, and deep ravines which are beyond the reach of the prairie fires, but they are not thick enough for cover, either for Indians or game, neither are they sufficiently tall or strong scaffolds of deposit for the Indian dead. These are wrapped in buffalo hides and either fastened among the branches of the tall cottonwood trees or placed on platforms on the open prairie, made of four tall stakes (about 15 feet high). There the bodies remain beyond the reach of the coyote until time heals the grief of relatives and [——] the dry and bleached bones upon the ground.

In the angle between the Powder and Yellowstone we determined to make the beginning of the survey back to the Missouri River. A large cottonwood post was cut and planted well in the ground, and in pencil we wrote on its clear white peeled surface a note to our delayed friends whom

[9]A few paragraphs later Rosser more accurately describes the Powder River that day as being low.

we expected to meet there, not knowing that their escort, under Col. Baker, had been defeated by the Indians near the mouth of Pryor Creek, a few days after leaving Fort Ellis and the survey had been abandoned.

After these preliminaries had been attended to, we went down to the bank of the river, with Gen. Stanley and his staff to clear our throats of alkaline dust with a glass of champagne. After the bottles had been emptied, they were refilled with dispatches to the leading St. Louis newspapers, then tightly re-corked and thrown in the river that they might float down into civilization, be picked up, and their contents published, thus carrying to our absent and anxious friends, tidings of our whereabouts and safety.

The clear rapid water of the Yellowstone pressed the inflowing muddy water of [the] Powder close into shore. As the river at this point was not very deep, one of the scouts rode across the red border, which the stained water of Powder gave to the pure and Yellowstone the canteens. Most of the water, even the springs found on the plains, is strongly alkaline and after drinking it for two months the water in the Yellowstone, clear and cold, just from the melted snow in the mountains was to us a most delicious beverage.

The supply brought by the scout was soon exhausted by our party under the cottonwood trees on the Powder and I rode upon the plateau in search of the scout to have the canteens replenished. As I came up out of the low ground, my attention was attracted to the bluffs about a mile off, whence the faint sound of muskets and shouting was heard. I saw a man dressed in dark clothes, carrying a gun which glistened in the sunlight as he ran (for he was running) and there were a dozen or more mounted Indians after him, but the ground was so gullied and broken that he had thus far baffled pursuit, yet the Indians were firing at him and riding rapidly ahead of him to cut off his retreat, and although he was making the best efforts with his legs and rifle, his chances of escape, unassisted, were gloomy. I yelled, "Indians! Indians!" with all the power of my lungs and in an instant every man was on his feet and every musket was manned!

The ponies of the scouts were unsaddled, yet before the echo of my alarm had died in the hills they were mounted and rapidly flying to the rescue! Seeing our scouts approaching, the hostiles gave up the chase and got out on smooth ground to meet the impending attack. I had never seen an Indian battle and I looked on with a most animated interest.

The hostiles were more numerous, but the scouts were better armed and as they approached each other, in the most apparent headlong manner, I fully expected to see the hair and scalps fly, but they did not close.

As soon as they got within range they scattered to the right and left, rode in circles, and fired without effect. Finally the scouts formed a line as skirmishers and advanced on the hostiles in the order of cavalry and drove them back into the bluffs. But when the scouts attempted to follow them into the rough ground, the hostiles dismounted and repulsed them. Then the hostiles would remount and ride in pursuit of the scouts, but when turned upon they would scamper back to the hills. A great deal of noise was made and many shots fired, but no scalps were lost or taken and no one hurt as far as [we were] informed upon the results of the "battle."

While this Indian circus was going on, our pedestrian friend, with dark clothes and glittering musket, bravely and breathlessly ran into our midst. His eyes were distended, his cheeks were pale, and he really had a very homesick and unhappy appearance—yet we recognized him as our jolly and witty rodman, Mr. H. C. Davis who, after getting over his temporary embarrassment, admitted that he had never before, upon a hunt, been so fortunate in finding game so abundant and large.[10] This episode thoroughly roused the command, the artillery was put into battery, the infantry was formed into line, and the engineers resumed work upon a new line of survey back to the Missouri River, making our cottonwood post the starting point.

We had moved but a short distance from the bank of Powder River when a tall Indian riding a sneaking, cowardly looking wolf of a pony appeared on the opposite bank of the river and asked for a conference. He was in full war regalia, wearing nothing but paint, breach clout and an old weather beaten stove-pipe silk hat, which gave him a most grotesque appearance.[11] This August prince of the plains was at once recognized as The Gaul, a minor chief of the Hunkpapa band of Sioux, distinguished for his cunning, cruelty and courage. After the usual complaint[12] that we had invaded <u>his</u> country with big guns that frightened away the deer and

[10]In a side note Rosser stated, "Davis is now the efficient and popular Assistant General Passenger Agent of the N.P.R.R." This dates the article to 1895 or '96.

[11]Sitting Bull wore a stove pipe hat in 1872—and no one else so describes Gall—so Rosser apparently confused the two.

[12][Footnote by Rosser.] "Our guides translated as quickly as the Indian spoke."

buffalo, he majestically ordered us to be gone, or he would bring down his people upon us and destroy us. Gen. Stanley informed him that his mission was one of peace, but he rudely disputed this avowal with "you lie," and rode back rapidly to the high hills in his rear.

After gesticulating a great deal and boasting of his prowess he cautiously rode back, charging Stanley with treachery and demanding that he send all the soldiers off that he might safely approach and talk to him. He declaimed on the courage and heroism of his people whom the white "dogs" had made poor, beating his breast said that his braves, their wives and their children were starving, that their ribs were as bare as their "*travois poles.*" But all the time he was thus addressing Gen. Stanley he was in pantomime giving instructions to his warriors who were engaged with our scouts. Our guide[13] suspected treachery and cautioned Gen. Stanley to withdraw, which he did, but not a moment too soon, for two individuals had cautiously and unseen crept along the bank of the Yellowstone and concealed themselves in the willows, and awaited the opportunity to give the General the contents of their rifles.

Gaul had withdrawn the greater part of his force and held them in reserve just over the hill in rear of him, to be called out should the opportune moment arrive. When Stanley withdrew they poured out, lined on the bank of the river and opened a vigorous fire on us. Fortunately we were beyond their range. They were more or less covered by the cottonwoods trees and we were on the open prairie. Their position was too low to depress the artillery so as to reach them, and not "spoiling for a fight" we proceeded with our work, and left them alone. After we were well out of the way, they swam across the river, danced around our cottonwood post, shot at our line stakes, and yelled defiantly at us, but did not pursue us.[14] We remained in the neighborhood several days and saw nothing more of the Indians but remembering the narrow escape of Mr. Davis none of us followed the antelope out [of] sight of camp.

We had finished on the Yellowstone and were retracing our line up O'Fallon's Creek. It was early in the morning, the valley was narrow with

[13]There were actually three guides with Stanley who spoke Sioux: Bottineau, Clement, and Louis Agard (1832–90), who had been raised by the Sioux and personally knew Gall. Trivia: Agard was the great-uncle of Henry Agard Wallace, vice president of the United States during Franklin Roosevelt's third term.

[14]To the Indians, the surveyors' pole and stakes were a form of powerful "magic," having the strength to draw locomotives to them, and thus they refused to touch them.

a deep canyon ahead of us and Gen. Stanley determined to keep along with us [versus separating for the day as was the usual case], with his entire command until we reached the open prairie. It was well that he took this precaution, for just as the head of the column was entering the canyon, The Gaul, reinforced by Sitting Bull, opened a brisk fire upon us. Fortunately they were unable to restrain their warriors and they fired too soon for their purposes. The wagon train was on open ground, and was at once parked [in a defensive position], which was done by closing the intervals between wagons, till the wheels almost or quite touched, and all faced to the center of a large circle making the wagons afford a protection to the animals. The troops retreated from the canyon when fired upon, and a line of skirmishers was deployed and pushed up into the hills for the purpose of clearing the canyon.

After this disposition of the troops had been made, and all were engaged except a company of infantry and a Gatling gun, composing the rear guard, about a hundred mounted Indians moved up the valley in our rear. The engineers were ordered to keep themselves inside the corral or park, the Indians came on at a gallop, and in the order of veterans. The Gatling gun was opened on them, and as soon as they came within the range of musketry, the infantry rear guard began firing. But on they came, and in perfect order.

I felt sure that they would charge on the infantry and make an effort to reach and destroy our train. The artillery was well served, but the infantry became very nervous and fired badly. The Indians maintained order and came boldly and rapidly on. There was no hope to secure and reinforce in time to meet this charge, and it appeared certain that they would ride into the guard. I jumped off my horse, took a stand near the Gatling gun, and began using my rifle vigorously. Their painted features and fiendish yells were indeed appalling, but our only safety was to fight, and there was at least one in the rear guard who loaded and fired as rapidly as he could,[15] yet our ranks began to waver and the officers had to threaten the men, to keep them in line and make them do their duty. The Indians came up within a hundred yards of us and a collision seemed inevitable, when suddenly, as one man, they broke by the flank and dashed up a ravine and out of sight into the hills! Oh, my! What relief, I believe if they had come on twenty-five feet further, that

[15]Rosser is referring to himself.

our infantry would have broken and run under the cover of the wagons. I then realized the difference between an army fighting for patriotism and one fighting for a business rather than work for a living. A company of volunteers in the armies of '61 & '65 would have killed every Indian in that platoon under similar circumstances, and these escaped without leaving a single one of their miserable carcasses behind.

The Indians in the hills, seeing that their plans had miscarried, made some loud threats about seeing us again,[16] and then remounted their ponies and disappeared over the prairie. On the prairie the voice of a white man is hoarse and indistinct, while that of the Indian sounds like a huge bell; loud, clear and voluminous. Their lungs equal those of the mighty giant Goliath, and every word they utter can be heard at a great distance.[17]

Surveyors' Diaries

Northern Pacific Surveyors' Field Diary, August 17–22

SUN, AUG. 18TH Clear—Started from camp as usual to run the chronometer line to Powder river—Had two teams and sent the line through flying—Ran the eight miles by ten-o'clock. Shortly after the line party started Gen Stanley and a large number of the officers with two hundred soldiers and the Napoleon gun followed—After reaching Powder river some of the officers opened some wine etc. and we were just beginning to enjoy ourselves when the alarm of Indians was sounded and this time it proved to be a reality—On reaching the top of the bank we saw a footman about a mile out on the plain and about a dozen mounted Indians on the full run after him and our scouts also going on the full run to the rescue which they fortunately accomplish[ed] as on their approach the Indians turned back and made for the hills where quite a lively skirmish occurred but no one was injured—The footman that was chased turned out to be no less a person than H. C. Davis of Eckelson's party—Shortly after the skirmish the Indians all crossed to the west

[16]Sitting Bull's "speech" was actually delivered before the firing began.

[17]Many frontier observers noted that Plains Indians were great runners, a reflection of having previously lived for thousands of years without horses. Like Indians in the Andes over the centuries, could some of them literally have developed larger chests and lung capacity, thus explaining Rosser's observation and one repeated about Sitting Bull in 1873?

side of Powder River and came down to the bank and talked with one of the scouts—As they turned to go away the Indians shot at our men when another skirmish hotter than the other one ensued and two Indians were shot—after this we were not troubled any more. Ran three miles of transit line. Reached camp at sundown.

MON. AUG. 19TH Clear & very hot—Ran 6½ miles of transit line—

TUES. AUG. 20TH Clear & very hot until four o'clock when it clouded up and the wind blew very hard—Camped. Moved two miles to the east went back to Sta. 235 and chained the line—Eckelson commenced at 545 and ran back until we met then my party took up the line at 545 and ran to 626—quit work at 4½.

WED. AUG. 21ST Clear & strong wind—both parties ran today—I ran from 626 to 741+ & from there on ran curves and Eckelson's party ran tangent—Indians appeared again and chased Eckelson and shot one of the scout's horses.

THURS. AUG. 22ND Clear & very warm—Indians showed themselves as soon as we left camp—Reynolds[18] the hunter for the government outfit drove [off] a dozen of them single handed—About nine the Indians reappeared in force and the line party had to withdraw to the train while the skirmish line was thrown out to drive them back and we returned to work again as soon as this was done. The skirmish lasted about an hour and half when the Indians retired—They numbered about one hundred & have evidently been heavily reinforced—Line was run from Sta 1029 to 1142.

Diary of Henry C. Davis, August 15–24

Henry Chandler Davis (1849–1910) was from Vermont, his family friendly with Gregory Smith's. After attending college in Wisconsin and working on the Mississippi River he joined the Northern Pacific in 1869. Davis claimed to be one of those driving the first spike at the 1870 groundbreaking and, more

[18]Charles A. "Lonesome Charlie" Reynolds (1842–76). This almost mythical hunter, frontiersman, and guide who died with Custer is deeply entrenched in Custer lore. He saw military service throughout the war and served in a similar capacity on the 1873 survey and 1874 Black Hills expedition. Well known before Custer, Sheridan hired him the previous winter when Russia's Grand Duke Alexis wanted to hunt buffalo.

certifiably, participated in 1883's final spike ceremonies. Intelligent and outgoing, Rosser named a creek (south of Medora) after him in 1871.

Davis's 1872 claim to "fame" was as the common-sense challenged surveyor who wandered off hunting for agates, only to be spotted by Gall. Somehow Davis was rescued; even more astonishing he kept a journal that he wrote in that night and, amazingly, the diary survived. Of all the handwritten documents reviewed for this book, the diary entry for August 18 was among the most difficult to read and transcribe.

AUGUST 15. [Indians] Struck camp at 5 O'clock—Had a very long crooked road plenty of dust. About noon we came upon a quite large stream flowing north 50° west. We camp at 5 O'clock on it and call it O'Fallon's Creek.

AUGUST 16. At about 3 O'clock—just before daybreak we were woke[n] by firing along the picket line and Indians yelling and we suddenly realized all the feelings—amusing and otherwise that are upon an Indian. 20 Indians came into camp just after we left. After a long dusty march we camp on O'Fallon's Creek seven miles from Yellowstone.

AUGUST 17. Broke camp at usual time. Made our way down creek quite slowly. About 9 O'clock scouts came in from all sides and reported Indians to the number of some hundreds—anywhere from 3[00] to 5[00].[19] The train was corralled and [the surveying] line party called in. Everything was placed on a fighting footing. We had no trouble however. An attempt was [made] during the afternoon to get them in and have a council with [them but] no success. We camp at 3 O'clock on the Yellowstone and have target practice with the Napoleon and [——]. I go to Powder River with line party tomorrow.

AUGUST 18.[20] Breakfast at 6 o'clock. I was not assigned to duty on line but Gen. Rosser asked me to accompany him to the mouth of the Powder River. Wishing to improve on every opportunity to collect [agate] specimens I started out ahead of [the] escort with the intention of following up the Yellowstone till I joined them.[21] After having gone about four miles

[19]The number was never above thirty.

[20]This virtually unbroken entry is edited with punctuation and paragraphs where none existed.

[21]Davis's incredible stupidity, especially after seeing Indians for two days, cannot be overemphasized.

and taken a good bath I heard the Napoleon gun firing some distance away. Seeing that I had missed in my calculations regarding the distance and direction of my objective point I made good time to the top of a Butte nearby to ascertain the exact location of the firing and the cause.

No sooner had I reached the top of the Butte than I saw off to my left—about three miles distant—in a very large coulee something like 150 horsemen riding in all directions and from every movement showing themselves to the Indians. Before coming to this conclusion I discussed every probability that could present itself to the contrary—was it possible that they could belong to our party? No because we had no such numbers of mounted men with us—Could it be that Col. Haydon and his escort under Col. Baker had got down from Deer Lodge[22] and that his Crow scouts were giving vent to their good feeling by the antics that [I] could plainly see them going through—not any—for they would be nearer the party which I thought to be in the vicinity from which I saw the smoke rise when the Napoleon was fired the last time.

Now it was that I had a lively discussion with myself regarding what direction it was best for me to take—whether toward camp four miles east of me and in going stand a chance of running into these Redskins as they might be making a dash at the stock that was picketed between me and the camp. I concluded to make for the point where I heard the firing and took the precaution to get down under the river bank and follow it as far as possible. Up to this time I don't think that the Indians saw me, but after I had walked about an hour, I came upon a very extensive piece of rapids[23] where the cut bank of the river came close to the water's edge. At this point I was obliged to go onto the bluffs and take my chances of being seen. When I looked for the point I was seeking to reach, it looked miles away when in reality it was scarcely one. Trusting to luck—counting a good deal on my legs [and] wishing that I was anywhere on earth but where [I] was, I struck out over hills and through coulees—after making about half a mile I reached the trail made but a short distance before by the escort. I followed it a short distance when my worst fears seemed in a fair way to be realized.

My situation was this[:] I could see no trace of the party save the trail I was on, and I was so near the place where the gun was fired that [I]

[22]Why Davis thought Deer Lodge and not Bozeman–Fort Ellis is unknown.

[23]Wolf Rapids, named by William Clark in 1806.

concluded they had moved after firing the last shot. I didn't know but what they might have gone back and I have missed in crossing some coulee. In the midst of my confusion I was startled by hearing the sound of running horses—the jingle [of] bells on their saddles, the well known war whoop—compared with which everything I ever heard from the lamentations of a Colored Baptist revival—to the noise an amateur gets out of a piano when practicing.

There I was, how far from help I knew not, with a band of bloodthirsty Sioux coming down on me like the wind. The imperfect expression of language forbids my telling all I felt; I remember that [I] told myself my time had come and that to run was useless for I could see no place to run to—I felt real deep seated home sickness—a terrible longing to be anywhere else—I thought about Longfellow the [——]. Yes, I had in my mind everything that can move fast. I heard men tell of being caught the same way and turning their thoughts back in repentance on their past lives and misdeeds and mistakes. I thought of but one and that the one that made me think I had with me an army rifle and 40 rounds of ammunition and coolly—lemme see—and coolly concluded to walk along at my regular gait and trust to luck and as soon as they showed themselves give them my 40 rounds as fast and sure as I knew how—and a good deal more that I don't remember.

All this ran through my head in a tenth of the time I take to tell it in [here]. I know [I] passed down through a deep canyon and as I rose on the further side my attention was attracted by the Scouts coming at a flying rate of speed to my rescue. It was then that I began to feel that I had a chance left and every minute—yes second—it seemed to increase for the scouts were coming at a fearful pace—I knew that I was safe when I saw the Redskins to the top of the hill about 500 feet behind me and seeing the scouts turn and run without firing. Soon the Scouts were between me and danger and I was quietly walking my way to [our] party which I discovered about a mile away—just at the junction of the Powder and Yellowstone.

When I reached them I found them watching the skirmish going on between the Scouts and Hostiles on the bluff near where I first saw them. A running fire was kept up for nearly an hour when the Scouts came in[,] the Indians having disappeared among the hills.

In less than a quarter of an hour they were seen about 30 in number—circling round to the north and crossing [the] Powder River [was] one of them. A medium sized fellow [Gall] seemed to be riding down toward us—he dismounted when a thousand yards distant and taking off his blanket approached us on the opposite side of the river giving signs of peace. He called loud for Louis Agard our interpreter who coming forward announced the infamous character of a Sioux renegade.

The Gaul wished to have a talk with Gen. Stanley. The General came forward and [had] a few minutes talk with the red rascal [and it] developed that he wanted us to leave what he called his country. He did not ask for but discussed rations and at the same time declared that he'd give the white man no more peace. Gen. Stanley flatly refused him rations it now became evident that an attempt was going to be made to assassinate the General for the Redskins seemed to be concentrating in the thicket close to the river as the General moved away the Scouts moved up to protect him.

The transit line was then started east and just as Reed the transit man was leaving the first Hub—Crack! came a volley from the thicket then followed a skirmish of about 20 minutes during which time at least three Indians were killed.[24] I claim the honor of firing the first salutary shot that sent one of them to the happy hunting ground. We drove them out of the bushes and then turned homeward.

When two miles away we could see them returning to recover their dead. I was obliged to ride home in an ambulance on account of blistered feet. Thus ends one of the most full days of my life.

AUGUST 19–23.[25]

AUGUST 24. I started east tonight.

Source: 1872 Henry Chandler Davis Journal, Microfilm no. 189, State Historical Society of Wisconsin, Madison.

Davis doesn't say it, but while running from the Sioux, his feet had been cut to pieces by rocks and cactus, and he was no longer capable of sustained walking. On August 24 some one hundred soldiers, Davis, Linsley, teamsters in

[24]In the fighting, one of the Hunkpapa was badly wounded but recovered. There is no collaboration to Davis's claim.

[25]Entries for August 19–23 are omitted to avoid repeating details found elsewhere in this chapter.

empty wagons, and others returned to Fort Rice. Two weeks later the highly motivated Davis joined the resupply wagons that were being sent back to the Stanley's column. However, after a few days he discovered that his feet hadn't healed sufficiently for him to effectively work. When the now empty wagons returned to Fort Rice in late September, Davis was with them.

Rosser Diary, August 16–24

AUGUST 16. With the Indian scouts. I rode in advance of the command down the creek to the Yellowstone. Saw no signs of Indians nor game of any description. The Indians who were with me reported that they saw a large bear and a few Indians, but I doubt their report. Indians attacked camp this A.M. but accomplished nothing.

AUGUST 17. Indian signal fires on the hills last night and when we were out, about an hour this morning, the scouts reported Indians advancing in large force over the hills and that they were very near us. The Commander was at once prepared for action but the report proved false. We camp tonight on the Yellowstone.

AUGUST 18. After a late breakfast we took an escort and the line party. General Stanley and mounted escort accompanying us. We ran our line to Powder River. Soon after arriving the Indians made their appearance and a skirmish ensued in which one Indian was killed. We began locating each . . . [Entry abruptly stops.]

MONDAY, AUGUST 19. Continued the line down the Yellowstone River but owing to the distance from the chronometer line I could only run preliminary lines. Will have to return tomorrow and finish the work. Commanding Officer shows signs of getting on a spree. Very hot and dry. Took a bath in the Yellowstone.

AUGUST 20. Returned and finished location where I had the preliminary [——] run yesterday. The run is a fair one but can be much improved upon careful examination. Very warm day and I am quite unwell and very tired. Moved camp two miles.

AUGUST 21. Moved camp back on O'Fallon Creek. Ran line down the Yellowstone and O'Fallon's Creek about 2½ miles. Indians appeared on the hills and came near capturing one of our scouts. Shot the scout's horse. D.C. L[insley] is expected to leave us tomorrow and all hands are busy writing letters home.

AUGUST 22. The Indians attacked us this A.M. after leaving camp with determination and vigor but after a skirmish of about 2 hours they withdrew and we moved on. D.C.L. won't leave until the Indians get quiet. We continue on location up O'Fallon's Creek. I can't bear to think of home.

AUGUST 23. Having good luck with our line, have got in the creek at several points but will not have to bridge. The line lies well and appearing to give satisfaction. I am dreadfully homesick and uneasy about my precious family. May God bless them.[26]

AUGUST 24. It is now decided that D.C.L. shall leave tonight. Worked hard till 2 P.M. then went to camp to fix up maps and write letters. Wrote Lizzy and sent her the leaves of my journal to date. Sincerely hope all will go well with her in my painful absence.

Stanley's Official Report, August 18–22

On the morning of the 18th I escorted the Engineers to the mouth of Powder River, eight miles west of our camp. Here we fired the 12 pd. [Napoleon] Gun several times, hoping to apprize Colonel Baker's command, expected daily from Montana, of our whereabouts. Whilst the Engineers were building their initial mound, and the command rested, an alarm of Indians was given, and 25 or 30 Indians were seen coming down the hills full speed after a foolish young man named Davis, belonging to the Engineer party, who had strayed off to hunt agates. Our scouts and all the mounted officers, most of whom had unsaddled, jumped upon their horses and made a race with the Indians, the prize being Mr. Davis's life. Our people outran the savages and after the exchange of a few shots, the Indians got off, and crossed to the west side of Powder River.

[T]hen their chief, the Gaul (Hunkpapa) came up to the bank of the river, laid down his arms, and called out that he wanted to talk with me. I laid down my pistol and moved forward to the opposite bank of the river, and proposed to meet him on the bare sand bar in the middle of the river. He declined doing this, and the talk that ensued amounted to very little. He wanted to know what we were doing, and when I told

[26]While Rosser had no way of knowing, his youngest child, one-year-old John Pelham, had just died. One doesn't have to believe in the paranormal to find this entry spooky.

him, wanted to know what we were going to pay, and finally threatened to bring all the bands and give us a big fight. As his young men continued to come over the hill and conceal themselves in the wood behind him, and as he could not or would not stop them, I thought it time to leave, and had hardly retired a few paces when the Indians fired on our party. Our men returned their fire, and drove them off. Feeling the folly of trying to catch them with our jaded horses, the Engineers commenced their location, and we moved off towards our camp. I learn since that our people wounded one Indian pretty badly who has since recovered.

It was determined to locate the railroad as far east as the head of Heart River upon the route we had marched over; this led us back up O'Fallon's Creek. On the evening of the 21st of August, just as we were going into camp 20 or 30 Indians concealed themselves behind the rough hills ¾ mile from camp, and five of them made a daring dash to cut off one of our Santee Scouts. As soon as this brave fellow saw that they were between him and our camp he rode straight at them, and passed through them. One of the Indians shot his horse in the jaw, and the horse fell, but immediately recovering, the Santee sprang on his back and escaped to camp. In an instant Lieut. Brush in command of the Scouts was out with his men, and pursued the Indians fully 15 miles at full speed before they gave up the chase.

Next morning the Indians hid themselves on a high and very rugged promontory that overlooked the valley of O'Fallon's Creek; a few Indians showed themselves, doubtless expecting to draw our scouts into another pursuit. I had however forbidden the scouts to pursue them again, as the horses of the scouts were badly used up, and the hostiles finding they could not draw the scouts and opened fire upon the escort with the Engineers. Two companies of the 8th Infantry, commanded by Captain Worth, and part of Captain Goodloe's company of the 22nd, soon drove the Indians beyond range, and proceeded with the work.[27] During the skirmish an Indian calling himself "Sitting Bull" stood behind a rook on a precipice, and addressed us at length, calling over the names of all the bands he would bring to our extermination; his list comprising pretty much all the Sioux, the Arapaho and the Cheyenne.

---

[27] Some sources cite Capt. Archibald H. Goodloe (1842–99) as being in charge of the artillery. From Kentucky; West Point '65. He resigned from the army in the early 1880s because of Bright's disease.

I had purposed sending a train back to Fort Rice for supplies as soon as we reached the Yellowstone, but as the train as well as the main force would have to retrace the road up O'Fallon's Creek, and as this route ran through cottonwood groves, and was commanded on both sides by steep and rugged hills, I kept the command together until the night of the 24th of August, when it being practicable for the train to clear the wooded valley by a night march I started Captain Read,[28] with two companies of the 8th Infantry, and one of the 17th Infantry, back to Fort Rice. Mr. D. C. Linsley, Assistant Chief Engineer, and Captain Kellogg, A.D.C. to the Lieut. General, accompanied this command.

Nibsey Swipes, Special Correspondent

Other than his existence, little is known about "Nibsey Swipes," the army's non-commissioned officer (my guess) or an officer who accompanied the 1872 survey. Under the pseudonym of Nibsey (also spelled "Nipsey") he sent numerous letters to the Sioux City Daily Journal, *all of which are reproduced here. Swipes was well known among his comrades, and one specific reference to him survives. To the degree that we know anything about him, it might be inferred that he was a moody, pessimistic man. Whoever Swipes was, he likely wasn't on the 1873 Yellowstone survey, or else he had been told not to submit any more letters.*

Sioux City Daily Journal, September 27, 1872

AMONG THE REDS
GENERAL STANLEY'S EXPEDITION TO THE YELLOWSTONE—
OPPOSITION OF THE INDIANS—
A LIVELY FIGHT NEAR YELLOWSTONE.

August 24

Editor Journal:

As the mail is about to start for Fort Rice in a few hours, I will endeavor to give an account of the progress of the expedition thus far.

---

[28]Capt. Edwin W. H. Read (d. 1875). Like many soldiers stationed in the South, Read was captured at the war's beginning and exchanged in 1862. He led the Eighth Infantry at Gettysburg as a captain. He was appointed second lieutenant in the mid-1850s. It is possible that he enlisted before the Mexican War.

Nothing of much importance transpired during the march up, save a few attempts made by the red skins to enter our corral and steal cattle, but a volley from the sentinels soon convinced them that they were discovered and that they hastily beat a retreat. These thieves who numbered about twenty dogged us all the way from the Heart River to the Yellowstone, watching our movements and any chance to become possessed of some of our horses, mules, or bulls, but nary a horse, mule, or bull did they get.

Our [outward] trip occupied twenty-three days, and when we arrived at the Yellowstone we were given a few days to rest, which we much needed. The Yellowstone is a beautiful stream and is delicious water to drink.

In the meantime the engineers started for Powder River, four miles from our camp on the Yellowstone, to survey in that direction. They were escorted by Companies K and B of the Eighth Infantry and A of the Seventeenth. Arriving at Powder River, they were met by about fifty Indians headed by the chief Big Gaul, who began to talk to General Stanley across the river. He wanted Stanley to understand that he could not cross the Powder River. Stanley asked him through the interpreter what he would do to in case he did cross the river. Big Gaul replied that he would drive him and his soldiers back. The scouts of the expedition then crossed and a sharp fight ensued. This was an interesting spectacle: both parties mounted and charging at each other with furious mettle. In a short time three of the Indians were unsaddled and the whole party retreated in haste across the hills. General Stanley told them in parleying with them that there would be a railroad through shortly in spite of their threats; to which they replied that they would all die preventing it, and that their country would never be troubled with anything of the kind. I may as well state here that the soldiers did not interfere in this fight at all, as the scouts had not enough to give them common amusement, as the head scout remarked.

We finally bid good bye to the friendly Yellowstone, and started on our return trip on the 20th inst. Up to the present we have not come a great way on our return as the engineers are operating slowly.

The Indians this morning paid us a visit and opened fire on our pickets on the bluffs, distant half a mile from camp, determined, no doubt, to have revenge for their fellow braves. The pickets, which were posted on this bluff since 3 o'clock this morning, were completely surrounded [in mid-morning] before they were aware of their situation. As soon as the pickets commenced firing the train was immediately corralled and the artillery put in position. Very soon after Company K, Eighth Infantry, commanded by Maj. W. S. Worth[29] and First Lieutenant F. A. Whitney,[30] deployed as skirmishers and double quicked up the bluff to the assistance of the little picket party, which in a few moments would have been gobbled up.[31] As the company ascended the bluffs the bullets whistled around their ears pretty lively, while they gained the point occupied by the pickets. As soon as the Indians saw the reinforcements they retreated about a half mile and signaled for more help. In a short time the bluffs around us were swarming with Indians, some mounted, more on foot, and all evidently preparing to make one grand charge on our little party, which was neatly deployed among the bluffs.

Very soon the red skins began to advance on us, and when within good range of us we opened up on them with our breech loaders. It was evident that they had no idea that our rifled would carry so far, and they sought the shelter of the bluffs. They made more than fifty attempts to flank us, but in each attempt they were foiled and after three hours smart firing, in which time there must have been over five hundred shots exchanged, they retreated, not however before telling us that "they would meet us again very soon and give us enough of it." The Indians numbered something near two-hundred warriors, and there are more, no doubt, laying back to give us another brush. We expect some more trouble from them very soon.

We have had no one killed or wounded thus far, thank goodness! and I believe the Indians have had quite enough at present. We have rubbed

---

[29]William S. Worth (1840–1904), actually a captain in 1872, was born Winfield Scott Worth. His father, William J., was a combat veteran of the War of 1812 and a highly regarded general in the Mexican War. He had been close to Scott for thirty-five years before they had a falling out in Mexico after which he changed his son's name. The younger Worth was a Civil War veteran and continued serving until he received a bad wound at San Juan Hill (1898) and was forced to retire.

[30]Folliot Augustus Whitney (1842–1900?), a Civil War cavalry veteran. The date of his death is uncertain, with one source listing 1900 and another stating that he did not retire until 1906.

[31]I have been unable to confirm this story (about the picket party), which appears to contradict other accounts.

out nearly half a dozen of the starved heathens already, and that is more to them than fifty to us in one sense. In today's skirmish Lieut. Ray, of the Eighth Infantry, plugged one of them who ventured rather a little too near. Lieut. R is a fine shot and kept us considerably in antelope and elk meat on the trip.[32]

As my space is limited, and as I have no desk on which to write except my knee, and no seat except that which I fetched from Ireland with me, I must end this letter. I will write again by the next mail and endeavor to give you a description of the country we pass through.

About thirty wagons are to start for Fort Rice immediately and [on their return] bring us rations. They will meet us about half way in, and they will be welcome, as we are beginning to feel the pangs of hunger already, thanks to our good Uncle Sam.

[Signed] Nibsey Swipes

Sioux City Daily Journal, October 17

THE YELLOWSTONE EXPEDITION

GATLING GUNS VS. INDIANS—THE PROGRESS MADE—SHORT OF RATIONS—THE TROOPS RAGGED AND WORN OUT.

Little Missouri River,[33] September 23, 1872

Editor Journal:

Your readers, no doubt, have heard a good deal about this expedition, by this time, and the attempts the Indians have made to protect their land from the white usurper. But the few attempts they have made are not sufficient argument against the barking Gatling gun and breech-loading rifles. When they attacked us on the 23rd of last month, and were defeated by about sixty of us, and driven helter-skelter across the bluffs, they thought that it would not be a very risky job to try and

[32]Lt. Philip (or Patrick) Henry Ray (1842–1911), a commissary officer, began the Civil War by enlisting (1861) and retired as a brigadier general (1907). Ray played a key role on the 1873 surveying expedition and was cited by both Custer and Stanley in their reports. He fought against Apaches, led American troops in Alaska during the gold rush, and fought Philippine "insurrectionists" in the early 1900s. Ray apparently left no papers, which is unfortunate, for he led as fascinating a life as anybody in the army during this period.

[33]Letter likely written somewhere near present-day Marmath, North Dakota.

capture our train and kill all our soldiers. As the events proved, they were disappointed, and, happily, not a soldier or white man was injured.

If Mr. Big Gaul ever again attacks any party crossing the plains, he will, in my opinion, first look sharply to see if they have got any Gatlings with them. Oh, they are wicked bull-dogs, and are just the thing to make treaties with the Indians. I am certain if Uncle Sam saw Big Gaul and company retreating to the cover of the bluff from the shower of grape which these Gatlings sent whistling around his red skin, he would send all Indian agents home to await orders. The Gatling is the thing; *it speaks to the point.* We expected another tussle with them ere this, but as they have felt of us before, I rather think they have given it up and gone to hunt up their winter supply of beef. We wish them good luck, as long as they keep away from us. They are troublesome customers, and running over bluffs and across ravines after mounted Indians "ain't fun, by a darned site," as a gawky private remarked; and he was right—was John Samuel.

We have made very slow progress thus far, and do not think we can reach Fort Rice until about the middle of November.[34] Jack Frost has already shown himself, and we cannot compel that gentleman to clear out as easily as we did Big Gaul, for *he* cares nothing about powder and lead. We are out sixty days on this expedition, and are likely to be sixty more. We are dirty and ragged; our old war hats have assumed a dozen shapes, and are sticking on our heads without rim or crown. We are almost shoeless, and, in fact, as bad as Falstaff's ragamuffins, who had but a shirt and half in the whole battalion. But if they would only give us enough "hardtack" to eat we would be satisfied. We don't get enough of that article to enable us to perform the hard duty which we have to do on this expedition. Weary marching, guards and pickets, fetching water and wood on our shoulder, sometimes three miles, after a hard day's march in the hot sun, with empty stomachs, and up at 3 o'clock in the morning, so as to be ready to receive the Indians in case they attack us at daybreak. Breakfast at daybreak. Ah, that breakfast! Save the mark. Strike tents immediately after; forward—Lord only knows where—wherever Old Ross pleases to point his horse's head.

They do not survey over eight miles per day on an average. Sometimes we lay over for several days at a camp, whenever they come to a bluff which they have some difficulty to find the level of, etc. Then, again, they

[34]Actually the expedition was back in Fort Rice three and a half weeks later.

make mistakes, which they have to go back, sometimes twenty miles or more, to rectify. They sometimes think they may find a better or a more direct and evener route ten or twelve miles to the right or left of us, and accordingly we have to follow the leader, Old Ross. General Stanley follows as meek as a lamb, with his poor, ragged worn out troops, who have about two small hard tacks and a half to last them for the day.[35]

The cold weather has begun to set in, and we are very poorly protected against it. We are very much afraid that before we reach Heart River, Mr. Thermometer will step down to twenty degrees below freeze me; and if he does, Lord help us.

While writing this I learned that the scouts have just reported fifty wagons coming out to meet us. They are loaded with provisions for us—good!—but not much clothing. They bring the mail, and I dare say a fresh supply of whiskey; but the latter we poor privates must not even dare to look at, lest we might get tipsy. My space being limited, I must close this letter, and I hope that General Stanley and General Rosser will be through in time for us to have our Christmas dinner in winter quarters somewhere.

[Signed] Nibsey Swipes

Sioux City Daily Journal, September 12

DAKOTA

SPECIAL DISPATCH TO THE *SIOUX CITY DAILY JOURNAL*

FORT SULLY. D.T., SEPT 11—A hostile Indian came into Cheyenne [River] Agency today, bringing a horse that was stolen from the agent a few days since. He says that Gen. Stanley, on command of the Yellowstone expedition has had no fight with Indians so far. A body of Indians, Minniconjous, Santees [an Eastern Sioux tribe], and Hunkpapas, headed

[35]Hardtack was a theoretically nutritious cracker but susceptible to mold, insects, and mice. Made of flour, water, and salt, it was long lasting, apparently had a disgusting taste, and was said to break teeth. While the above comment on hardtack likely represents the norm in hardtack speculation—admittedly a small cadre—frontier historian Paul Hedren writes (and corrects) me as follows: "A fresh hardtack cracker was hardly disgusting and only the most aged, infected, or poorly made broke teeth, and even then ever so rarely. They are still made today by a bakery [G. H. Bent Co. Crackers] dating to before the Civil War that was an actual purveyor to the Union army and, as a former living historian, I've eaten hundreds of these. . . . They taste like oversize saltines. Yum."

by Sitting Bull, Gaul, the Neck,[36] and Racer, made one or two feints, but not taking the troops unawares as was expected, they withdrew, neither side losing any men. The Indian reports that these bands have dispersed and formed hunting parties, giving up all further intentions of opposing the survey. The information is deemed reliable.[37]

[36]"Neck" possibly refers to No Neck. There were two Lakotas of that name. One of these—likely the one in the article—was a Hunkpapa war chief, possibly older than Sitting Bull, who joined him in 1877's flight into Canada. He appears to have died in the mid-1880s. I am unable to identify Racer.

[37]In fact, this was one of the most accurate of all the contemporary newspaper stories.

Western Survey

Major Barlow's Journal, August 12–20

August 14.—This morning about 3 o'clock the command was startled by the report of several rifle-shots, and in a few moments a lively firing was opened by our sentinels along the line facing the slough. It was soon evident that the Indians were making a night attack. The guard at once turned out and commenced driving in the mules that were out "on herd." The principal attack seemed to come from the right, where Captain Rawn's[38] battalion of infantry soon deployed and poured several volleys into the thicket, in which the Indians were massed in considerable force. These Indians were at length driven from their cover, and were seen scampering across the open field and getting beyond the range of our guns as rapidly as possible. In the meantime the mules were driven in by the guard and safely corralled within the enclosure formed by the wagons.

On the left a small squad of citizens, who had been traveling under our protection, were in camp. The Indians fell upon them early in the fight, and succeeded in capturing a fine rifle and some ammunition from one of the tents. Near this point an Indian was shot by one of the citizens, and afterwards killed by a soldier of Captain Thompson's company. This Indian's body was dragged within our line. A number of Indian horses were killed during the fight; and it is quite probable that several Indians shared the same fate, as their bodies were seen being carried off from the field.

The bluffs above were lined with savages, who, together with those

[38]While Barlow's report uses the name Bacon, there was no such officer in the Seventh Infantry. I have substituted Rawn's name, as he led the four companies of infantry.

nearer at hand, kept up a continual firing upon our camp. Bullets penetrated our tents, struck three or four men, killed two horses on the picket-line, and wounded several mules. During all this time it was too dark to estimate the force of the Indians, or even conjecture what their plans might be. As soon as it became sufficiently light to render distant objects visible, the plain to our right, beyond the swamp, was seen swarming with red-skins charging in circles, and keeping up such a yelling as only they are capable of.

Every available point upon the bluffs in our front for a mile in extent was occupied by them, while from the various ravines and other places of ambush came rifle-balls often sent with most unpleasant precision. The whole force was variously estimated at from five hundred to one thousand.

Captain Ball, of the Second Cavalry, with his company dismounted, went out on the left, and drove the enemy from the bluffs in that direction.

Lieutenant Grugan,[39] Second Cavalry, with fifteen men of his company, mounted, went part way up the bluff, directly in front. These two forces, not acting in concert, returned after accomplishing their immediate object.

The Indians' headquarters seemed to have been established upon a prominent bluff on the left of their line, from whence couriers were frequently dispatched. This point and other bluffs in its vicinity were densely crowded with Indians during the fight, and remained in their possession for several hours after daylight. The firing ceased about 7 o'clock, and, though the fight was over, apprehension was felt that a force of the enemy might cross the river, and, from the brush on the opposite shore and intervening islands, seriously annoy us and force us either to abandon our position upon the river-bank or sacrifice a number of animals. But they had not the sagacity to attempt this, and were undoubtedly satisfied to get off without further loss to their own party. They left two dead bodies, and ten or twelve dead ponies. Our casualties were as follows: One sergeant of the Seventh Infantry, killed, shot through the head; one private, same regiment, wounded in leg; Private Ward, Second Cavalry, supposed mortally wounded, shot through eye and head; Private Cox, same regiment, driver of headquarters wagon, also thought to be mortally wounded, shot through the bowels; one of the

[39]Frank C. Grugan (1842–1917) enlisted in a Pennsylvania regiment during the Civil War, was in the Second Cavalry for almost a dozen years, transferred to the artillery, and participated in the Spanish-American War, apparently seeing fighting in Cuba.

citizens mortally wounded, shot through upper part of body and spine. Eight animals were stolen, two horses killed, and all the beef-cattle, about fifteen in number, driven off. Toward noon a reconnaissance was made by Captain Ball, with his company mounted, which revealed the fact that the Indians had all left our immediate vicinity.

About 2 P.M. we broke camp, the engineers resumed their survey for three miles, and the command moved down the river and encamped on its bank, four miles below the fighting-ground, in a large grove of cottonwood trees. Considerable apprehension was felt of another attack, as this position was favorable. The river, however, afforded some protection, but to prevent surprise the cavalry companies moved out beyond the brush, where they passed the night. The following morning's observation of the ground discovered the fact that the Indians had reconnoitered our position in large force, the soil being much trampled by ponies. But our precautions had doubtless intimidated the savages, and the command enjoyed the undisturbed rest which they so much needed.

AUGUST 15.—The surveyors went out under the escort of one company of cavalry, and the main column moved up the hill upon the level bench some ninety feet above the river bottom. This is a broad, level expanse, nearly barren, being stony, and yielding little vegetation except sage and prickly-pear. The surveyors make good progress, being only detained by small, sharp ravines which extend down from the mountains to the river. Over five miles of survey were made today. We encamped on the point of a bluff, between two ravines, and have an excellent defensible position. Just below us the river makes a broad bend, inclosing a considerable flat of bottom-land on this side, covered with good grass, and containing a grove of cottonwood trees. The animals are grazing in the bottom, and water is brought up to camp from the river. Across the stream a vast and nearly level meadow extends away several miles in breadth, gently ascending toward a low ridge in the distance. There is nothing particularly inviting in these landscapes except their extent, and even this feature becomes monotonous.

One or two Indian ponies were found on the trail of the Indians. This trail indicates that a greater part of the band which attacked us have crossed the river and gone southward.

AUGUST 16.—The camp was not moved today, the engineers wishing to take time to survey across the deep ravine just beyond our situation, and obtain some triangulations for topography. The hill on which we are camped is covered with pebbles of various kinds of mineral, onyx, chalcedony, marble, granite, &c.

AUGUST 17.—The command moved across the creek just below camp this morning. The table land here opens out a mile in breadth, but is very barren, being alkali soil, and yielding almost nothing except an occasional sage-plant and a few bunches of cactus. Skirting the western and northern limit is a steep bluff of rock fringed with stunted pines. These bluffs extend toward the river, and a few miles farther on completely bar our progress, along the bench near the stream. It will be necessary to break through and ascend these bluffs and continue our journey along their heights for a number of miles. Water and grass will probably be scarce through the entire region. As the movement could not be commenced today, camp was made on the edge of the bench overlooking the river and the vast bottom-lands upon the opposite side. It seems unfortunate that we cannot cross and continue our route on that side, as we could thus save much both in distance and time. From this camp "Pompey's Pillar," about fifteen miles distant, is visible.

Four large trails have been discovered crossing the river near this point, and remains of crude rafts have been seen upon which dead and wounded Indians were probably crossed over. Bloody clothes and other indications show that the savages received some injuries from our heavy firing. The remains of nearly all the beef-cattle have been found. The Indians appear to have had a very large force in this vicinity. Several large camping-places have been discovered, and the estimate of the number of Indians at the time of the fight was, perhaps, too small.

AUGUST 18.—We were startled last night by two false alarms, each of which brought the whole force out. The first alarm occurred at 10 P.M. and was occasioned by a shot from one of our sentinels, discharged at a floating log in the river, which he had mistaken for an Indian. About 3 A.M. another alarm arose from the fact that a sentinel went out of camp, and being seen beyond the line, was fired at by another of the guard. Nothing serious happened. No movement has been made today. The question has arisen regarding the propriety of our farther advance, in view of possible Indian hostilities. Colonel Haydon, in charge of the survey, presented to Colonel Baker a series of questions as to his opinion and that of other officers of the command, concerning the adequate strength of the escort, and its ability to protect the engineers when separated from the main column, and concerning the facilities for taking care of a large additional number of wounded in the event of another engagement. Colonel Baker replied that it is the unanimous opinion of himself and his officers that the escort can conduct the survey to Powder River and back in safety, and protect the engineers in their duty, and that there is ample provision for any wounded we are likely to have.

AUGUST 19.—Camp was not moved today, but the survey was extended. The day has been intensely hot, with a fierce wind blowing during the afternoon, which made camp-life exceedingly uncomfortable. Specimens of petrified fish were found in the black-clay rock, so abundant in the cliffs along the stream. This cliff is one hundred and twenty feet in height, and made up of clay, sand, and black rock. Colonel Haydon, in charge of survey, came in about 6 o'clock. He was able to make but two and a half miles of survey on the line, owing to the roughness of the country and the distance of the work from camp, his party being compelled to walk six miles before beginning their labors, and to return the same distance after finishing for the day.

AUGUST 20.—The difficulties under which Colonel Haydon labors, the impossibility of continuing work without dividing the force, determined him last night to notify Colonel Baker that he would not operate any farther on the present line, but that he desires to go across to the Musselshell, and survey back to Fort Ellis by that line. This is probably a just and wise determination, in consideration of all the circumstances of the case.

In accordance with Colonel Haydon's wishes, camp was moved this morning at 7 o'clock, and the column started for the Musselshell River.

Our course as advised by Mitch [Boyer], the guide, has been in a northwest direction, over a dry, alkali country, nearly barren from lack of moisture. Much difficulty was encountered in crossing the numerous sharp gullies and ridges, but the teams were fresh, and, with the occasional help of shovels and picks, got along very well, making about eight miles. We camped on a small ravine containing muddy water in pools, probably rain-water, and very strongly impregnated with alkali. This spot is not agreeable, as the grass is very scarce, the principal productions being prickly-pear and sage-brush, with some grease wood for variety. Soon after pitching the tents a terrible gale arose and filled the air with such a storm of dust that there was no relief either within or without the tents, and besides nearly every tent in camp was quickly blown down. All attempts at cooking, reading, or sleeping were of necessity abandoned.

Baker's Leadership and Surveyor Activities, August 12–20, 1872

Date	Baker and Main Column	Surveyor Progress & Escort Size
Aug. 12	Reach Yellowstone River and establish camp.	Surveyors with main column.
Aug. 13	In camp.	Surveyors pick up 1871 survey line (7 mi. west of camp) and run it east toward campsite.
Aug. 14	Camp attacked about 3 A.M. Break camp about 2 P.M. and march 3 mi. to new camp.	4–5 mi. surveyed, moving parallel to Baker column. Escort is one cavalry company, approximately 40 men.
Aug. 15	March 2½ mi. to new camp.	5 mi. surveyed (2½ mi. beyond camp). Escort again one cavalry company.
Aug. 16	Camp does not move. "Sharp ravines extend down from the mountains to the [Yellowstone]."	1½ to 2 mi. surveyed, ending 4-plus mi. from camp. Escort, one cavalry company.
Aug. 17	Camp moves ¼ mi. to cross a creek. "Bluffs extend [to] river."	Limited surveying near camp. No escort information given. Pompey's Pillar seen.
Baker decides not to cross Yellowstone River.		
Aug. 18	Camp does not move. False Indian alarms (waking entire camp) at 10 P.M. and 3 A.M.	Survey 2 mi., finish 6 mi. from camp; hot and humid. No escort information given. Haydon asks Baker about safety.
Aug. 19	Camp does not move. "Intensely hot with a fierce wind blowing all day."	Survey 2½ mi., reach 8½ mi. from camp. No escort information given. Haydon, exhausted, returns, signing letter that Baker gives him asking for the column to turn back.
Aug. 20	Camp moves 8 mi.	—

Source: 1872 Report of Major John W. Barlow, 42nd Congress, 3rd Session, U.S. Senate, Ex Doc No. 16. All mileage figures are approximations.

U.S. Army and Northern Pacific Communications following Baker's Fight, August 19–27

Col. John Gibbon to Gen. Winfield Scott Hancock in St. Paul, Department Headquarters, District of Montana (via telegram), Fort Ellis, Montana Territory, August 19, 1872

Sir,[40] The courier from Maj. Baker's camp conveying the intelligence of the Indian attack on his command reached here yesterday a few hours after my return to the post. The information from reliable officers of both cavalry and infantry is of such a character regarding Maj. Baker's condition at the commencement of the attack (2:30 am) and during its continuance for several hours, that I consider Capt. Rawn,[41] the next officer in rank, would have been fully justified in assuming command of the expedition, and *were there any means now for me to communicate with it, I should place him in command and arrest Maj. Baker* [emphasis added].

The small party which accompanied the courier (Ordinance Sergeant O'Niel [*sic*] of this post) left the camp at night and were fortunate in reaching the post, a distance of about 180 miles in less than four days unmolested. One of the officers states that *all the whiskey in camp had, since the fight, been spilled and from this fact I trust that no further like trouble will occur* [emphasis added].

When the troops once commenced to act they repulsed the Indians with but little difficulty. The command is simply able to take care of itself, and the warning it has had will be of advantage to it. I regret the necessary *ex parte* character of this statement, but deem it my duty to make it . . .

[Signed Col. John Gibbon]

Source: National Archives Record Group 393, Records of U.S. Army Continental Commands, Military Division of the Missouri, Letters Received, no. 2940 (1872).

The same day that Gibbon sent his telegram to St. Paul, it was forwarded to Sheridan in Chicago. Sheridan then sent a telegram to the War Department, omitting all references to Baker's "problems."

---

[40]Technically sent to Maj. O. D. Greene, acting adjutant general of Hancock's staff.

[41]Charles C. Rawn (1837–87). Rawn, a Civil War enlistee in 1861, was soon transferred to the Seventh U.S. Infantry where he was to spend most of his career.

Philip Sheridan, Chicago, to Brig. Gen. E. D. Townsend,[42] War Department, Washington, D.C., August 19, 1872

The camp of the Surveying Party of the Northern Pacific Railroad from Fort Ellis, Montana, under the command of Major Baker, was attacked by Sioux, Arapaho and Cheyenne numbering from four to eight hundred, at three o'clock on the morning on the morning of the fourteenth of August at the mouth of Pryor's Creek, Yellowstone Valley.[43] The attack continued for three hours when the Indians mentioned [retreated], taking with them all the beef cattle of its command. Baker lost five men and eight horses. The march of the surveying party was resumed, but much apprehension is felt of additional trouble when the expedition pushes down the valley.

[Signed] P. H. Sheridan

Source: National Archives, Record Group 94, Records of the Adjutant General's Office, Letters Received, #3323, 1872.

Sheridan to Brig. Gen. E. D. Townsend, War Department, August 20, 1872

[Sir]

Major Baker lost only one man killed, three soldiers and some citizens badly wounded: two horses killed and sixteen head of beef cattle and four mules captured. Indians killed unknown. Several ponies killed and captured. Major Baker has 382 men, with which the surveying party, employees, etc., makes his party I think *about five hundred strong* [emphasis added]. He has gone on, confining himself to the object of the expedition, mainly the protection of the surveying party. I think that most of the Indians attacking were from the vicinity of Fort Fetterman.[44]

[Signed] P. H. Sheridan

---

[42]Edward Davis Townsend (1817–93), West Point '37 (some sources state Harvard, although he did attend Boston's famed Latin School) was the army's adjutant general. Prior to the Mexican War he was assigned to the adjutant general's offices, remaining until his 1880 retirement. Townsend apparently died when he accidentally touched the live wire of a street trolley.

[43]Pryor Creek is on the opposite or southern side of the Yellowstone. At this time no one—Gibbon, Hancock, nor Sheridan—seemed aware that Baker had been unable to cross the Yellowstone.

[44]Why Sheridan included this misleading comment is unknown.

Source: Record Group 393, Military Division of the Missouri, Letters Sent, vol. 42.

Hancock to Commander, Fort Rice

St. Paul, Minn., August 19, 1862
Col. T. L. Crittenden, Commanding Officer, Fort Rice
To be sent [by currier] via Fort Cross—Jim River[45]

Major Baker in command of the Montana escort down the Yellowstone was attacked on the morning of the fourteenth instant at the mouth of Prices [*sic*] River by a War party of four or five hundred Sioux, Arapahos and Cheyenne. Indians were repulsed. Their loss not known. *Our loss is considerable* (emphasis added).[46] If you can do so, communicate this to Stanley by an Indian courier at once, if not send him by the first opportunity that offers. Subsequent to the fight Baker resumed his march. The Department commander [Hancock] desires you to communicate this information to the posts in the District and to Fort Buford.

By command of Major General Hancock.

Gregory Smith to Hancock, New York, August 22

In many respects the following letter from Gregory Smith is one of the summer's most interesting documents. In it Smith forwards telegrams that he has received from Helena office manager Joseph W. Flenniken, in which the latter extensively quotes from a clearly rattled John Haydon, the lead surveyor. The passages that quote Haydon are confusing in the extreme, devoid of punctuation and correct spelling. Faced with a choice of leaving the inherently unreadable original as is or attempting to "clean it up," I have chosen the latter—as best possible—if not always successfully.

[Sir:] The following telegram was just received at this office. "Yellowstone Party were attacked at Pryor's Creek August Fourteenth by four hundred Arapahoe Indians. One citizen wounded, fourteen beef cattle and five head of stock lost. Signed G. W. [*sic*] Flenniken, assistant engineer."

[45]"Jim River" was the James River; Fort Cross (changed to Fort Seward late in 1872) was in or immediately adjacent to western Jamestown, North Dakota, 110 miles east of Fort Rice. It was likely at the temporary end of the telegraph line to Bismarck.

[46]Where this incredible misinformation came from is not known.

At [a] later hour: the following just received. "Haydon reports Baker has called for two hundred Crows for defense. Haydon considered this injudicious. Opinion of officers [is] that one thousand troops necessary to get below Bighorn River. Troops demoralized. Viall[47] gone to Bitter Root informed him of state of affairs. Signed Flenniken"

This morning we have the following from Haydon. "Attacked at 3 A.M. by Northern Arapahos and Cheyenne. Attack strong, persistent and brave. This may be a beginning with their strongest force. May be an effort to feel our strength [and] if be true we can only be worn down by repeated attacks. If further attack be made on bluffs below we may be cut off in detail. My party with a strong escort together in camp fifteen miles distance.[48] Infantry and Cavalry fought well. Military mistake in not putting forth full energy at right time when Cavalry should have acted. There appears lack of command. I think escort inadequate for duty expected. Many officers think we could [not] go below Bighorn without one thousand men. Excessive guard duty will wear the men out.

Later. Colonel Baker has issued orders to send for two hundred Crows to assist in defense, harass the enemy and steal horses. I think this a mistake. [I] expect an attack hourly. If you can influence the assignment of larger force do it at once and we can finish survey in thirty days. Signed H. J. [*sic*] Haydon."

These reports indicate trouble far more serious than we had anticipated. It is doubtful if the survey from Fort Ellis eastward down the Yellowstone to meet survey west from the Missouri can be prosecuted unless additional troops are sent forward immediately. Of the wisdom of exciting the Crows to fight the other bands, you can better judge than I. It is possible to order more troops [to] sustain our party. Please reply and give us the necessary protection at once. . . .

Signed J. Gregory Smith

Source: National Archives Record Group 393, Records of U.S. Army Continental Commands, Military Division of the Missouri, Letters Received, no. 2940.

---

[47]Col. J. A. Viall was the Montana superintendent of Indian affairs.

[48]This sentence is particularly difficult to understand. Did Haydon write this a few days after the attack when he moved forward to survey, but Baker refused to move the camp?

The most interesting aspect of Hancock's letter is how quickly senior army officers "circled the wagons" in regard to "outside" criticism. Hancock's putdown of Smith, Flenniken, and Haydon could not be more thorough. Hancock's letter was the first step in setting up the Northern Pacific as scapegoat and Haydon's subsequent public humiliation.

Hancock to Smith, August 23

Hon. J. Gregory Smith,

I received from Major Baker an account of the attack on his party. Its tone was calm. He repulsed the attack and has resumed his march. The dispatch to you from Engineer Flenniken gives the same information as Baker's dispatch to me only less in detail. Flenniken does not seem excited and I think the dispatches signed Haydon must be unreliable. Is Haydon with the party or does he telegraph from some point in the rear? Major Baker says nothing of employing Crows nor does he ask for more troops.[49] Until I hear further from Baker or other reliable sources I shall presume he is strong enough, is master of the situation, and will accomplish the object of his expedition.

I do not at all believe the troops are demoralized and believe that Baker exercised judgment in not pursuing the Indians after repulsing them. If he had done so it would, necessarily, have divided his force compelling him to fight a force at least equal to half his men—the Cavalry, where by maintaining a defensive attitude, although still advancing into their country he keeps his whole force in hand ready to fight at any time, and does not endanger the main objective of the expedition, unless the strength of the Indians becomes so great as to overwhelm his entire force, which is not probable.

There are no more troops available in the Department for Baker's reinforcement in case he shall require it, unless Colonel Stanley unites with them. If there had been, Baker's Command would have been made larger at the outset. It would take thirty days to get reinforcements to him of any such strength as you indicate, one thousand men, either troops which might be sent from other Departments or Montana Volunteers and, as at present advised I see no necessity for such action.

Any information in regard to the fight must have come in with the party that brought Baker's dispatches for me and I therefore think there

[49] Nor has any collaborating article or letter(s) supporting Haydon's assertions been found.

is no good reason to suppose there is anything in the situation of which Baker did not advise me.

[Etc., from Hancock's office]

Source: National Archives, Record Group No. 393, Records of the U.S. Army Continental Commands, Military Division of the Missouri, Letters Received, # 2763 (1872). No paragraphs in original.

Smith to Hancock, Boston, August 27, 1872

Hancock's response led to the following reply, which is not just a classic example of very rapid backpeddling but equally instructive concerning how to withdraw from controversy while simultaneously protecting yourself. (Also known as CYA *in the military.)*

Thank you for your telegram. It is more than probable that Haydon may have telegraphed under some excitement. I have confidence in your superior judgment, and shall consider the party safe so long as you regard it as so. Haydon is with the party and I understand was with them at the time of the attack.

Signed J. Gregory Smith

Source: National Archives, Record Group 393, Records of U.S. Army Continental Commands, Military Division of the Missouri, Letters Received, # 2763 (1872).

Newspaper Accounts

Little is more astonishing than the speed and reasonable accuracy with which the Helena Herald *and the* Bozeman Avant Courier *reported the August 14 fight. The following stories fully capture the drama and concern following Baker's fight and its aftermath.*

Completely missing from all early accounts was that the expedition was on the wrong *side of the Yellowstone. Most articles mention the fighting occurring at Pryor's Creek, which enters the river from its eastern/southern side. Newspaper editors would have had Montana maps and therefore assumed that Baker had crossed the Yellowstone as planned. Other articles discuss the*

Indian "victory," the army blaming the surveyors, an editorial from the New York Herald, *and the final sad document—surveyor John Haydon's pathetic letter to the* Helena Herald *after he realized he had been tricked by Baker and had become the focus of blame for the survey's failure.*

What most of the articles have in common—and which has been modified—is a lack of paragraphs and the misspelling of Haydon's name and Pryor's Creek.

Helena Daily Herald, August 19

FROM THE LOWER YELLOWSTONE
FIGHT BETWEEN THE MILITARY AND INDIANS.
ONE SOLDIER KILLED AND SEVERAL WOUNDED

The Northern Pacific Railroad surveyors in the Lower Yellowstone country are, it appears, being somewhat interrupted by the hostile attitude of the Indians. An attack upon the military escort of the surveying party on the 14th inst. resulted in the killing of one soldier, and the wounding of three others and a citizen is reported. Pryor's Creek, the scene of the fight, is a tributary stream of the Yellowstone, some 130 miles below the Crow reservation of Fort Parker. The following dispatch was received at the Indian Office last evening.

Pryor's Creek, August 14th.
Via Bozeman, August 18th.

Col. J. A. Viall,
[Montana] Superintendent of Indian Affairs

We were attacked this morning by 400 Arapaho Indians. One soldier killed and three badly wounded. One citizen also badly wounded. Also lost fourteen beef steers and five head of United States stock [horses or mules].

Helena Daily Herald, August 20

PARTICULARS OF THE ENGAGEMENT
SERG'T MCLAREN KILLED—LIST OF WOUNDED—
SPLENDID FIGHTING BY THE INFANTRY—
GALLANT PURSUIT OF THE RED DEVILS
BY THE CAVALRY, ETC.

FROM OUR SPECIAL CORRESPONDENT.
YELLOWSTONE RIVER, AUGUST 15TH, 1872

TO THE EDITOR OF THE *HERALD*:

Yesterday morning, about half past two o'clock, our camp was attacked by a strong force of Cheyenne and Arapaho Indians. We were poorly posted in the open space, surrounded by heavy timber and thick underbrush. The Indian tactics were excellent. Making a demonstration upon the center of our camp, the major portion of their force was posted in thick underbrush and directed against our right flank, which rested on the river. Here they actually approached, unobserved, within a hundred yards of our camp. "E," "G," "D," and "I" of the 7th infantry, respectively commanded by Capt's Rawn, Browning, and Lieut's Reed and Quinton,[50] were deployed as skirmishers and moved forward to drive the Indians from the timber—a task gallantly performed by the men under a severe fire from the Indians. It was fortunate that the Indians fired wildly. We occupied the timber with the loss of one man killed outright (Serg't McCarran [*sic*], "C" Company, 7th Infantry), and four wounded, two of them mortally.

The Indians succeeded in getting away with five mules, one horse, and sixteen head of cattle. After being driven from the wood the Indians posted themselves upon a high ridge [that angled back] from 300 to 1,000 yards in the rear of the timber, which latter point they occupied for some hours, keeping up a desultory fire upon our position. The loss of the Indians is unknown, although it is supposed to be severe, from the fact that we found twelve dead horses in our front. "E" company fired three volleys into thick underbrush immediately in their front, which was

[50]Capt. George Leslie Browning (ca. 1840–82), New York enlistee at the beginning of the Civil War. Likely in the military when he died. William Quinton was born in Ireland (ca. 1841), graduated from a Chicago high school in 1860, enlisted at the beginning of the war, and retired as a brigadier general in 1902.

literally swarming with Indians, and as Mr. Lo was seen to swarm out in a hurry, it is supposed that he found the place uncomfortably warm.

[Signed] PEREGRINE

Helena Daily Herald, August 20

Following is a second letter to the Herald, *published alongside the one above. Neither* Herald *correspondent can be identified today.*

YELLOWSTONE EXPEDITION, AUGUST 15, 1872

To the Editor of the *Herald*

Yesterday morning, the 14th inst., at 2 o'clock we were attacked by a war party of about six or eight hundred Sioux and Arapaho Indians. They surrounded the camp and fired into it from every side, with horrid whoops and yells, endeavoring to stampede the horses and mules. The troops sprung to arms at once, and advanced in a skirmish line through the darkness toward a belt of timber in which the Indians were posted. Their fire was at first quite heavy and regular, but the infantry companies—"C" "E" "G" and "D" of the 7th—pouring into them two volleys in rapid succession, advanced at double-quick and drove them from their shelter to some high hills about a half mile from the camp.

These Indians seemed to be well armed with repeating rifles and revolvers, and fought daringly. Soon after reaching the hills, seeing the cavalry preparing to advance, they retreated in haste down the river. The cavalry followed them about five miles and then returned.

Our loss was Serg't McClellan [*sic*], Co. "C," 7th killed, private O'Malley, Co. "E," wounded in the leg; privates Carl and Ward of the cavalry wounded in the head and body supposed to be mortally; and a citizen prospector, named Francis, dangerously wounded through the body. We also had one horse killed and five mules stampeded.

The Indians lost, we think, about ten or twelve killed, but with the exception of two we failed to get the bodies, and we cannot say with certainty. Several were seen to fall from their horses, but were at once picked up by their companions and carried off. We found seven of their horses killed by our shot. We find that they have crossed the river, and as our large [wagon] train impedes pursuit, they are, I regret to say, beyond our reach.

Col. Haydon began his survey the day before the attack, and as it evidently does not meet with the approbation of the Lo family, I think we shall hear from them soon again.

The officers and enlisted men of the command behaved handsomely in the fight, and all are eager for another call from our red brethren.

A Serg't and escort start this evening for Fort Ellis, and by him I send this letter.

[Signed] G. L. B.

Bozeman Avant Courier, August 22

THE YELLOWSTONE EXPEDITION
ATTACK ON THE N.P.R.R. SURVEYING PARTY.
ONE WHITE MAN KILLED AND FOUR WOUNDED.
THE FIGHT LASTS SEVERAL HOURS
THE SAVAGES REPULSED! THIRTY INDIANS KILLED!

The military expedition which left this place about three weeks since as an escort to the Northern Pacific Railroad surveying party was attacked on the morning of the 14th inst. at the mouth of Pryor's Creek, about 160 miles from Bozeman, by a band of Sioux Indians, variously estimated from 400 to 700 strong. The party was encamped in the flat near the trees, and the attack was made at about half past 2 o'clock while most of the men were asleep.

As soon as the alarm was given the men were roused to action, and the fight commenced and lasted to about 9 o'clock A.M., when the Indians retreated. Capt. Ball[51] says the fight was one of the most determined he ever witnessed with the Indians, who fought with dogged determination, and made several attempts to break into the camp, but were repulsed on every attempt. The loss on the part of the expedition is: one killed and four wounded, viz: Serg't. McClaran, Co. C., shot through the head and instantly killed; Private Ward, Co. L., 2nd Cav., shot through the eye, the shot coming out at the ear; Private Cox, Co. F., 2nd Cav., shot through the bowels; a private of the 7th Inft., name unknown, wounded in the leg; and an attache of the surveying party, named Francis, shot in the back.

The Indian loss is estimated at about 30, but as, in accordance with Indian tactics, the killed and wounded were carried off as fast as they fell, it is impossible to give correct figures. One of the chiefs was killed

[51]Capt. Edward Ball (ca. 1824–84), Second Cavalry. Ball enlisted in the army in 1846 (under the name David Ray) and served in the Mexican, Civil, and various Indian wars. Exceptionally honest, in early 1876 Ball reported he had been offered a $1,000 bribe to ignore shoddy goods being sent to the Crow Indian reservation. Perhaps predictably, because a leading Montana citizen was involved (Nelson Story), a grand jury refused to indict.

and his body captured before the Indians could get away with it, and after the fight, thirty-two Indian horses were found killed on the field, and numerous trophies left behind by the savages in their hasty retreat, were secured by the troops.

Thus early has the campaign opened, and the Indians have commenced their opposition in the advancement and construction of the Northern Pacific Railroad, and from the reports we have received in their conduct in the late fight, it is apparent they will prove to be a formidable foe to contend with. They have commenced operations even earlier than expected; it was not generally supposed that any hostile demonstrations would be made until the party had crossed the Big Horn River, and probably the force engaged is but a small portion of that that can be brought to operate against the expedition in a very short time.

Col. Baker's reputation as an Indian fighter is well established; his coolness and courage have more than once been tested; but the mere handful of men composing his command are totally inadequate to the task of proceeding through a country infested with hostile savages, thoroughly organized, with an avowed determination to resist the accomplishment of the work which the expedition he protects has in hand. One thing is patent—if the small force of Indians engaged in the late contest—in numbers not exceeding 700—are able to make a seven hours fight against troops but little inferior to themselves, numerically, they certainly can surround and annihilate them by a concentration of [their] forces. And this accomplished, what is to prevent them their victorious career through the very heart of our settlements[?]

This [Gallatin] valley is now left almost entirely unprotected; the troops are absent, and will probably have all they want to do to protect themselves; a numerous and savage foe is known to be within a hundred miles of this place, and it is necessary that some steps be immediately taken to guard against the impending danger.

There are over eight-hundred needle guns[52] with ammunition in

[52]A needle gun was a rifle with a needlelike firing pin that could strike through a paper cartridge and, later, a metal cartridge. The gun itself was usually rifled (versus smooth bore) and a vast improvement over the muskets still in common use during the Civil War. Numerous inventors/tinkerers developed a wide variety of needle guns, with varying degrees of practical battlefield success. Austria, for example, lost its 1866 war with Prussia in large part because its newly designed needle gun proved unreliable under combat conditions. Western historian Paul Hutton writes, "The term was altogether common on the frontier and

Virginia City; let our citizens take immediate steps to procure a supply at this place [Bozeman], so that in case of an invasion we may not be entirely unprotected to defend our homes against the savage incursions, and let other measures be adopted to insure the general safety. "One ounce of prevention is worth a pound of cure," so let action be taken at once.

Bozeman Avant Courier, September 5

ABANDONED

Advices from the Yellowstone bring us the unwelcome intelligence that the Northern Pacific Railroad surveying party which left here a few weeks since to complete the survey of the Yellowstone country have been compelled to abandon their task for this season for a lack of sufficient force to protect the surveyors during its prosecution.

We are informed that after the fight at Pryor's Creek, the camp moved down the valley about four miles, when a council was held, and *to Col. Haydon's inquiry as to whether the military commandant* [*sic*] *had sufficient men to protect the surveyors at their work as well as the supply train, Col. Baker replied that he had not* [emphasis added]; therefore Col. Haydon concluded that it would not be practicable to carry on operations with so small a force in the very heart of Indian country and turned back, and the party are now engaged on the survey of the Musselshell.

[Line unintelligible] work could not be successfully consummated this season, as it will probably be the cause of several months delay in the establishment and construction of the railroad to this point, and will retard to a serious extent the settlement and development of the rich valleys in that section of the country; besides being a heavy drawback to our prospects of speedy communication with the East by rail.

We are inclined to believe that the parties conducting the expedition have done the best they could under the circumstances, but it shows a great lack of foresight on the part of those in authority to send a party, having such important objects in view, on such an expedition without adequate protection, knowing as they most unquestionably did the hostile attitude of the Indians, through whose country it would have to operate, and that the savages had already declared their intention to contest the advancement of any such expedition through their domain.

---

referred simply to Springfield breechloading rifles and carbines, .45 and .50 calibers, which had a four or five inch long needle-like firing pin running through its breechblock."

But regrets are useless now, and we can only hope that the lesson they have learned this season will be a warning for the future, and that abundant means will be provided next year for the successful carrying on of the line of survey. That the road must and will be built is an established fact; that it will be hotly contested by the Indian tribes through whose country it will pass is equally certain; and while regretting the failure which has attended the operations of the surveyors this season, we can only patiently abide the time when they will be successful, and reiterate the hope that we expressed in these columns four months ago, "that efficient measures be inaugurated and sufficient force be kept on duty to effectively deter the savages from any attempt to plunder, murder, or in any way obstruct the progress of this great work," as well as to inflict summary punishment on them for their depredations.

New York Times, September 10

The following communiqué from General Sheridan's office was based on a story from the notoriously inaccurate Sioux City Daily Journal. *Why it was taken at face value by Sheridan's staff and telegraphed around the country is a mystery. What makes this particularly so is Sheridan's being hundreds of miles away in Minnesota—traveling to Dakota—while his staff was fully aware of Baker's limited casualties. Since it was clear the story would badly damage the Northern Pacific, the question needs to be asked why was it released or who permitted its release? Assuming that the story was not a snafu—given the army perhaps not a safe assumption—two options come to the suspicious mind: first, that somebody on Sheridan's staff was "reached," or second, that the army was making sure it wouldn't be blamed for Baker's failure.*

THE INDIANS
ATTACK ON MAJOR BAKER'S FORCE IN THE YELLOWSTONE VALLEY—FORTY SOLDIERS KILLED AND WOUNDED—SITTING BULL THREATENS TO GIVE GEN. STANLEY "PLENTY OF FIGHT."

CHICAGO, ILL.—A special from Sioux City says that part of the band of Indians who were engaged in the fight with Major Baker's forces in the Yellowstone Valley came into the Cheyenne [River] Agency a few days ago and reported that the fight was very severe, and that the troops had lost heavily, over forty being killed and wounded, while the Indian loss was only eleven. They report that Baker is so badly crippled that he is obliged to return to Fort Ellis without having accomplished the object of the expedition.

Gen. Stanley sent a detachment back to Fort Rice a few days since for supplies. They report that just prior to their departure Sitting Bull sent in a flag of truce and asked for a parley with Gen. Stanley which was granted. At the interview Sitting Bull declared that if Gen. Stanley did not turn back at once he would give him plenty of fight before he got back to the Missouri River.

It is estimated that there are now in the Yellowstone Valley from 15,000 to 20,000 hostile Warriors [emphasis added], and serious trouble is apprehended. Many of these Indians are Cheyenne and Arapaho from the Platte Valley [in Nebraska], with a large band of Ogallala Sioux.

Helena Daily Herald, September 11

THE HAYDON EXPEDITION

Deputy U.S. Marshall Hard who arrived a few days since, gave us some items of interest concerning the Haydon surveying party. The expedition has abandoned the survey of the Yellowstone Valley for the present, and are now engaged in surveying the Musselshell route. Upon reaching the first belt of Bad Lands, about 50 miles below the scene of the fight with the Sioux, it became necessary for the escort to divide, a part to go around by way of the Musselshell, and the other part to accompany the surveyors.[53]

On account of high water it was found impossible to cross the Yellowstone River with the [wagon] train and animals, and go down on the other side.[54] For these reasons Mr. Haydon decided to return. They crossed over to the Musselshell Valley over a rolling divide, crossing no mountain range, but found a broken country especially in contiguity to the Bad Lands. The country presents no obstacles to a railroad. The valley of the Musselshell looks very [promising for a railroad].

[53]This is incorrect; other than messengers, the entire party stayed together until reaching the upper portion of the Musselshell, one hundred miles to the north and somewhat to the east of Bozeman.

[54]This is the first mention that Baker-Haydon never crossed the Yellowstone.

Chicago Inter-Ocean, September 11

The following article appeared throughout the country on September 11. As can be seen, the Army's intent was to blame the Northern Pacific. To what extent did this reflect election year presidential politics? Certainly at this point the election was considered exceptionally close, and blaming the Northern Pacific was the "smart" move for the Grant administration

REPORTED FAILURE OF YELLOWSTONE EXPEDITION

WASHINGTON, SEPT. 10—The War Department has received a dispatch from General Sheridan dated Duluth, Minn., saying that Colonel Baker, after reaching the mountains of the Yellowstone Valley, was obliged to come back, on account of the surveying engineers deciding to go no further, for fear of the Indians, and therefore the expedition is a failure.

New York Herald, September 12

WHAT IS TO BE DONE WITH THE INDIANS?

All our recent dispatches from the West relative to the attitude and action of the Indians encourage the belief that at no distant day another Indian war will be forced upon us. The surveyors of the Northern Pacific Railroad are experiencing more than inconveniences in the accomplishment of their task. From [around the country] . . . the reports are more or less the same; and we are left little room to doubt that the rising of the Indians on a large scale is not only contemplated, but almost certain to take place.

From one section we learn that if the surveying expedition party attempts to return east the Indians are resolved to fight to the bitter end. From another quarter we learn that there has been severe fighting; that the troops have been outnumbered and badly beaten, and that in the Yellowstone Valley there are not fewer than twenty thousand hostile Indians well armed and ready for the fray.

So it is; so it has always been. Civilization approaches; the Indians rebel. They can make promises. They can take gifts. Of gift taking, in fact, they are never tired. We cannot promise them too much. We cannot give them too much. But in spite of all we do or can do they remain what they have ever been—the natural enemies of civilization. The hunting grounds are diminishing, the buffalo is scarce, the reservation arrangements do not satisfy, and, of course, the white man is the cause of all their trouble.

What is to be done? Are Indians to be allowed to stand in the way of the development of the country? Because the Indians object, are our Pacific railroad enterprises to be abandoned? Of course not. We say be just to and keep faith with the Indian. Let the government look sharply after the Indian agents. Much trouble has come from this source. The agents employed by government do not always do their duty. Let them be punished, and punished severely, when they are found unfaithful. Let us do everything just and fair by the Indian; but let us not allow him to stand in our way.

If he will persist in resisting authority, the cure is simple. Let a plentiful supply of troops be sent to the disaffected quarters; let there be no lack of ammunition; let Phil Sheridan take charge, and let him understand that the work is to be done promptly and effectually. Indian outrages on American citizens [the surveyors] in the discharge of national duty must cease. If they do not behave we must resort to the practices of ancient Israel and root them out.

Is the following first-person account true? The fact is that the body of this article involves only the unidentified writer, and there is no collaboration. Compare this with the battle-hardened Rosser's and tenderfoot Davis's accounts. This writer knows how he would vote, but you, the reader, decide for yourself.

New York Times, October 1, 1872

A BRUSH WITH INDIANS

SURPRISED BY ENEMY—A DASH FOR THE TIMBER—THE VALUE OF A REPEATING RIFLE.

The following letter, from a gentleman in the surveying party of the Northern Pacific Railroad and Yellowstone expedition, gives a lively picture of life on the Plains. The expedition was compelled to turn back, owing to superior forces of hostile Indians:

"Swimming Woman" Creek[55]
Valley of the Musselshell
Sunday, September 1, 1872

Having been cut off from all communication with the world outside of camp, I have been unable to write to you. But thanks to hostile Sioux,

[55]Unidentified, but given the date, this is near Roundup, Montana.

Arapahoe, Cheyenne, etc. we are slowly returning to civilization. You have doubtless heard of our battles, so I will confine myself to my own private affairs. Since reaching Pryor Creek, where 2,600 of them jumped our camp (800 strong), they have showed us every attention, sometimes serenading us two or three times a night. We seldom see them in the day-time. Once, however, I did get a "glimpse" of them. Thusly: Mr. Ingram [now unidentifiable] and I started out for a hunt, [we] had not gone far when we came upon a trail showing a single pony track. We thought it would be fun to find him and "lift his top knot [scalp]" simply as an evidence of affection.

Following on several miles, we saw a herd of buffaloes, and concluded to let "Mr. Lo" go until a more convenient season. Away we went for buffalo. The herd divided, I taking after one part and my friend the other. You know it does not take long to follow buffalo several miles. But wait a moment; here is other game. Skimming along the base of a hill, I saw eight Indians coming full tilt. No time to lose. About four or five miles away was a belt of timber. If I could reach that I should be fixed. I did not feel like giving battle against such odds. On every hand I saw a vast alkali plain. I must make that timber. All this took but an instant. My horse was foaming from the recent chase, but it was run or die, and I ran.

They had the advantage of fresh ponies and were partially between me and the timber already. Of course they saw my plan, and kept well to the left to cut me off. My horse flew like a bird, but the rascals had the inside track, and their war whoops came nearer and nearer until a volley told me that I was actually within range. I threw my old rifle back and sent them my compliments, which must have passed near, as they slackened for a moment. Another volley and a whoop was the answer, and they came rattling on as fast as ever. My spurs fairly plowed furrows in the poor animal's side as we flew along. My coat dropped from the saddle, but I did not want it. I was quite warm enough without a coat.

They were cutting me off from the timber, sure. I must change my course. Holding further to the right, I might reach a point, about a mile further away. This would lessen their advantage in point of distance. But my horse was giving out. I dashed the long Mexican spurs in the poor fellow's side and he only answered with a groan. Do you know how far a mile is under these circumstances? I was not scared, but I fully appreciated the situation.

Again I fired, but they did not heed it. My courage almost failed me as I saw them cutting me off from this last point. Neither spur or the iron ramrod of my rifle could quicken the pace of the poor foaming and panting brute. My last hope was failing me now. I think I felt my scalp getting a little loose about that time. Just as I had formed the desperate resolution of dismounting and fighting it out, I passed the head of a deep gully, such as you have seen in alkali districts. For them it would be impassable. They must make a detour to the right. This put me beyond range, and brought them directly in my rear. Five minutes more, my old "Cayus," (*anglice*, pony,) and we shall be out of harm's way. They are too cowardly to chase me far into the timber—besides I can give them a taste of my quality from behind one of those trees while they are in open fields. Those were very pleasing thoughts, but the reality was—they were gaining on me rapidly.

That war whoop was much nearer than I would wish, and the whistling bullets informed me that I was in range once more. A moment more and I gained the timber. Throwing myself from my horse, I took deliberate aim and fired. I could not repress a good, square, defiant Sioux war whoop as I saw one pony wounded. Again and again I gave it to them, while *they* were getting out of range. I had not been at all excited before, but the tables were so completely turned that I could no more help sending a war whoop after each shot than I could help firing, to see those eight cowardly villains getting out of the way of my old rifle! It was just beautiful!

Seeing them draw off upon a ridge out of range as if holding a council, I concluded I had satisfaction, so I "folded my tent like an Arab, and as silently" slid off into the timber, leading the poor old brute who had saved my scalp back to camp. The old fellow "fares sumptuously every day" now, as I told one of my servants I would relieve him from all other duty, so that he could take good care of the horse.

You will hear before this reaches you that theYellowstone Expedition and N.P.R.R. survey is a failure. They demonstrated the fact that our force was wholly inadequate to pass down the Yellowstone Valley. One morning the red rascals put four bullets into my tent, and four into my mess chest which stood in the cook tent. Our tents are nicely riddled, you may be sure. We held a council of war about two weeks ago, and concluded to abandon the expedition. We then crossed a spur of Bull Mountain, striking the Musselshell near the Great Bend, and since that time have been slowly ascending this river.

Helena Daily Herald, September 25

YELLOWSTONE SURVEY FROM ASS'T ENGINEER HAYDON

Camp Baker: South Fork Musselshell,[56]
Montana, September 17, 1872

Editor Daily Herald:

Your issue of the 11th of September under the caption of "Washington Intelligence" contains the following paragraph: "The War Department has received a dispatch from General Sheridan dated Duluth and staying that Colonel Baker, after reaching the mountains of the Yellowstone Valley, was obliged to come back on account of the surveying engineers deciding to go no further for fear of Indians, and therefore the expedition is a failure."

This is utterly untrue and Col. E. M. Baker, commanding the military escort, assures me that he made no report authorizing in any manner such a statement.

Very truly yours, J. A. Haydon, Assistant Engineer, Yellowstone Survey[57]

James H. Bradley's Account

Lt. James H. Bradley (1844–77), Seventh Infantry, is most remembered for his posthumous memoirs, March of the Montana Column. *Enlisting in the Union army before he was seventeen, Bradley survived numerous major battles and Confederate captivity, ending the war as a noncommissioned officer. Promoted to first lieutenant in 1866, he participated in the 1876 campaign under Col. Gibbon's command, kept a journal, and was one of the first who discovered the physical remains of Custer's command on June 27, 1876.*[58] *That winter Bradley*

[56]Likely somewhat to the west of Martinsdale, Montana, a small community on the now abandoned Chicago, Milwaukee, St. Paul & Pacific Railroad.

[57]This sad letter was the last date from which information concerning John Haydon can be found. He does not appear in the 1880 census, nor does his native state of Tennessee have any data on him after 1872.

[58]Col. John Gibbon (1827–96), West Point '47, was an aggressive, highly respected officer who fought in Mexico, against the Seminoles, quickly became a major general during the Civil War, performed gallantly at Gettysburg (where he was severely wounded, leading Plains Indians to call him, "One who limps." He was one of those chosen to be at the McLean house for Lee's surrender.

wrote a detailed account of that campaign, but shortly after the manuscript was completed, he was killed in action at Big Hole during the Nez Perce campaign.

In addition to the '76 campaign, Bradley's manuscript included a detailed description of the August 14, 1872, fight and Baker's inebriation. While Bradley was not on the western survey, a number of his fellow officers were, and there is little doubt that he consulted with them before his work was completed.[59] March of the Montana Column *is gracefully crafted and deservedly a classic; Bradley's candor and military insight add to the book's usefulness and enjoyment. In reality, Bradley's access to many of the participants, while the fighting was still fresh in their minds, makes him a primary source. This is the only entry in this book that has been published recently and is still in print.*

[In July] two surveying parties were to take the field—one to begin at the Missouri River and extend its explorations westward, the other on the upper Yellowstone and work down that stream till it should meet the eastern corps at the mouth of Powder River. The former was provided with an escort of nearly one thousand men commanded by Colonel David S. Stanley, 22nd Infantry, while to Colonel John Gibbon, 7th Infantry, commanding the District of Montana, was assigned the duty of providing from the troops of his command a suitable force for the protection of the western corps.

For this purpose Companies C, E, G, and I, 7th Infantry, were drawn from Fort Shaw [Montana], and companies F, G, H and L, 2nd Cavalry, from Fort Ellis, the whole force, which numbered about four hundred men, being placed under command of Major Eugene M. Baker, 2nd Cavalry. Having marched from their respective posts, they were all assembled at Shield's River on the thirtieth of July, 1872, and, being there joined by Colonel Haydon with his corps of surveyors, began their march down the Yellowstone the following day.

In the meantime a heavy force of Sioux warriors, variously estimated at from eight hundred to one thousand strong, were ascending the river upon a hostile incursion against the Crows; and about the twelfth of August discovered through their scouts that they were in the presence of Baker's command. This unexpected recontre [*sic*] created a division in

[59]A number of sources, including Dan Thrapp's *Encyclopedia of Frontier Biography,* assume Bradley participated at the 1872 Baker battle; however, Mark H. Brown's history of the area, *Plainsmen of the Yellowstone* (201), specifically states that he did not.

their councils, many being anxious to give over their former design and measure powers with the troops, while the more prudent minority were disposed to avoid so hazardous an enterprise and continue their advance on the less prepared and unsuspecting Crows.[60] At length, however, tempted by the large spoils in horses which they hoped by dexterous management to secure at little cost to themselves, they declared in favor of an attack upon the troops, and fixed upon the morning of August fourteenth for carrying the plan into effect.

The troops had now reached and were encamped upon the ground that became the scene of the fight. A party of surveyors escorted by a force of cavalry commanded by Captain Ball had the previous year carried the survey down the Yellowstone valley to the Place of Skulls, and the command having by easy marches[61] reached the field of the summer's work were resting in camp while Col. Haydon completed his arrangements for taking up and continuing the survey. The presence in the neighborhood of two or three Indian dogs had excited some apprehension that there were Indians about, but the general feeling was of confidence and security; and *not only were no especial precautions taken by the commander of the force to guard against an attack, but upon the very night fixed for it he permitted himself to become unfitted for the proper performance of his duties by an overindulgence in strong drink* [emphasis added].

The camp was pitched upon ground favorable for defense, being located upon the margin of the stream, with a timbered slough sweeping in a semi-circular direction around it so as to form in connection with the river what may be termed an island of two or three score acres area, the whole at long rifle-range from the adjacent bluffs. To have rendered the position wholly secure, however, it would have been necessary so to guard the slough that it could not be occupied by the enemy as a preliminary to their attack; but this was not done. Fortunately it was rather the purpose of the Indians to get possession of the animals of the command with as little fighting as possible to gain any decisive advantage over the troops, and their plans were laid accordingly.

Several hundred warriors were posted close on the lower side of the camp, where they were wholly screened from view by the timber and

[60] The Crows were aware of a possible attack, the reason why none joined Baker.

[61] If there were any reason to believe that Bradley was a participant, his mention of "easy marches" should dispel those thoughts.

willows growing in profusion at the lower extremity of the slough, while the remainder of their force was to seek by an attack upon the landward side of the camp to draw the troops in that direction, when the ambushed swarms would burst from their concealment, sweep over the camp, cut loose the horses, throw the troops into confusion by attacking their rear, and at the worst escape with the herd. With such caution and success did they, under cover of the darkness, reconnoiter the camp previous to the attack that they were enabled to steal several saddles out of the tents of a party of prospectors, who had joined the command, while their owners lay within them asleep, cut from their lines and make off with six mules picketed near the tent of the commanding officer, and kill a dog that threatened to betray their presence in the camp.

The guard that night consisted of a mixed detail of cavalry and infantry numbering twenty-six men, commanded by Lieutenant (now Captain) William Logan,[62] 7th Infantry, a brave and sagacious officer, who was of the number that suspected the presence of the Sioux and, having a premonition of the attack, did all that a vigilant officer of the guard could do to avert surprise. His guard was posted on the flank of the camp, away from the river and some three hundred yards distant there from, his sentinels covering the camp as far as possible, while the herds of beef cattle and mules of the government and contractor's trains, which had been left out to graze, and held well under cover of the guard upon the island already described, with a squad of herders over them to prevent straggling or stampede. The horses of the cavalry were tied at the picket lines within the limits of the camp.

Toward 3 o'clock A.M., of August fourteenth, the officer of the guard made the round of his sentinels and found all quiet, the guard tents and the timber growing along the slough. Soon afterward from the timber at different points along the landward side of the slough the Indians opened fire and advanced upon the island to attempt the capture of the herd. In a moment the boldest of them were mingled with the animals, but the few men posted over the herd stood their ground manfully, opening

[62]Logan (1832–77) was killed by a Nez Perce woman on the Big Hole battlefield, where Bradley also died. Born in an Anglican family in Ireland, after immigrating to New Orleans he enlisted in the Seventh Infantry and, according to some accounts, fought in the Mexican War. The oldest of twelve, he was directly descended from Scotland's legendary Robert de Bruce. He became a captain in 1874.

a rapid fire upon their assailants at close range, and at the same time endeavoring to put the herd in motion toward the corral. The guard was instantly under arms, by judicious management the animals were driven gently to the rear, the Sioux who had sought to stampede them being forced by the fire of the guard to fall back.[63]

A few moments sufficed to enable Lieutenant Logan to throw his entire guard between the Sioux and the herd, where deployed as skirmishers and lying down in the long grass the men opened fire upon the moving forms dimly seen before them through the gloom. After the first volley the Sioux maintained a scattering fire, but the unexpectedly hot reception given them by the guard soon caused them to retire from the timber to the open ground beyond, and within a few moments after the attack began, the ground was cleared of them and their fire had subsided into a few straggling shots.

Meantime the herders conducted the animals to the rear, where without confusion they were driven into the corral and rendered secure, none having been lost, except fifteen head of beef cattle which stubbornly refused to move with the herd and fell into the hands of the Sioux. When the firing began, the citizen prospectors, some twenty in number, seized their arms and took an advanced position on the left of the guard, where with Lieutenant Jacobs[64] at their head they took cover and opened battle on their individual account. The Sioux speedily recovered from their first repulse and returned to the attack, reoccupying the timber and appearing in considerable numbers on the open ground in front of the guard. But the citizens with Lieutenant Jacobs poured in a rapid fire upon their flank while the guard received them firmly in front handling their breech-loaders with such effect that again the Indians speedily withdrew.

---

[63]Later in his journal, Bradley writes: "To show how our gentlemen are amusing themselves I quote the journal of one of them: 'Up all night playing "Pedro" for eggs.' When the attack was made on Baker's command in 1872 it found several gentlemen wide awake and absorbed in the mysteries of 'poker,' from which circumstance the fight is sometimes jocularly called 'The Battle of Poker Flat.' Being already dressed, they were quickly at their posts." Bradley, *March of the Montana Column*, 129.

[64]Joshua West Jacobs (1843–1905) was born in Kentucky to a pro-Union family. He enlisted in 1861, gradually rising in rank until his 1904 retirement as a brigadier general. During the Spanish-American War, he contracted malaria in Cuba from which he never fully recovered. A late-in-life photo shows a thin Colonel Sanders look-alike complete with a white Van Dyke beard and flowing mustache, his hat set at a jaunty angle.

At the first alarm the troops had promptly formed in their company streets and awaited the orders of the officer in command. As soon as the infantry battalion was under arms, *Captain Rawn, its commander, reported to Major Baker for orders and found him still in bed, stupefied with drink, skeptical as to the presence of an enemy, and inclined to treat the whole alarm as a groundless fright upon the part of the guard. It was difficult to get any order from him* [emphasis added], but at last he directed Captain to Rawn to hold his men in camp; and, disgusted and angry, that officer returned to his command and upon his own responsibility deployed Companies E (Lieut. Reed) and G (Capt. Browning) in line on the lower side of the camp, facing the thicket in which the ambuscade had been formed. Lieutenant Reed occupied the right, with his right flank resting on the stream, and thus posted, the men of both companies lay down in the tall grass. As bullets were flying freely through the camp, the remainder of the command was ordered to lie down in their company streets.

Captain Thompson,[65] officer of the day, had gone to the front to ascertain the cause of alarm, and nearly lost his life by advancing recklessly too far beyond the guard. Finding the attack real, he so reported to the commanding officer, and a re-enforcement of about thirty cavalrymen under Lieutenant Hamilton[66] was sent forward to Logan's support. Captain Rawn at last received tardy orders to deploy his command, and thereupon placed Company C (Lieut. Quinton) in position on the left of the line already formed, and his own company, I on the left of C, and then by extending intervals to the left caused the four infantry companies to cover about half the front, the citizens and cavalry continuing the

[65]Lewis Thompson (1839–76) also enlisted in 1861 and, through family connections, quickly became a second lieutenant. Able and well liked, he saw service in the nasty Kansas-Missouri border fighting, then joined the Second Cavalry only to be captured at Gettysburg. He was exchanged a year later, again through family contacts. By 1876, he was suffering severe emotional problems that became significantly worse when he was one of the first (June 27) to discover the horrific, now-stinking remains of Custer's command. Three weeks later, he became badly drunk on cheap liquor (possibly poisoning himself) and while in excruciating agony shot himself in the chest.

[66]Samuel Todd Hamilton enlisted as a private in 1862 and retired as a captain in 1892. He was known to be living in Washington, D.C., at the turn of the century. In 1869, when engineer Milnor Roberts was exploring the feasibility of a northern transcontinental route for Jay Cooke, he visited Fort Ellis. Hamilton led a contingent of troops to escort him across the Bozeman Pass and into the Yellowstone Valley; there everybody spent a few bucolic days trout fishing.

line to the left till it enveloped the camp. This deployment was effected within about half an hour after the beginning of the attack.

As yet the Indians ambushed on the lower side of the camp had not betrayed their presence by a sound. It was now growing light, and seeing the movement of the troops toward the point of attack, but ignorant that while it was yet dark two whole companies had taken position directly in their front, they imagined that their stratagem had succeeded, and the way was open to the picket lines where the horses were tied.

They began therefore to make their way cautiously forward, but ere they emerged into view Lieutenant Reed discovered the movement in the sudden rustling and swaying of the willows in his front, and promptly swept the covert with his fire, pouring three volleys by company into the timber with the regularity and precision of the drill-ground. This sudden and unlooked for storm of bullets full in their faces filled the Indians with astonishment and dismay. Without waiting to return a shot, they swarmed from the timber like bees and spurred their horses away for the bluffs in headlong flight. As they passed the remainder of the line, Companies C, G, and I also opened fire and completed their utter discomfiture. True to the Indian custom they carried with them their killed and wounded slung across their horses in their front, and Lieutenant Quinton, who occupied a favorable position for observing their movements, counted eighteen thus borne from the field.

The slough with its timber belt was now in possession of the troops, and afforded them excellent cover from the desultory fire which the Indians maintained for the next three or four hours from the bluffs. After the repulse of the latter from their ambuscade, they attempted no movement of consequence, but remained for the most part gathered in crowds upon the distant bluffs. Occasionally some of the bolder warriors careened on horseback at full speed along the line; a few of whom were wounded for their pains. About 8 o'clock A.M., they suddenly disappeared, and a reconnaissance by Captain Ball's company of cavalry showed them in full retreat. The officers generally were eager to follow them, and *Major Baker at one time ordered Captain Rawn to get two of his companies in readiness to move, announcing his determination to take them and two companies of his "busters"—as he was pleased to call the cavalry—and pursue; but he soon forgot all about it or changed his mind* [emphasis added].

Accounts agree in representing the Indians as greatly demoralized.

They afterwards admitted a loss of over forty killed (all but three of whom they carried from the field), and there was also a large number of wounded, probably nearly a hundred in all. They fled with great precipitation, and marked their line of retreat with abandoned effects that would have impeded their flight. A vigorous pursuit, encumbered with dead and wounded as they were, would speedily have brought them to bay, *and an energetic commander with his wits about him and such a force at his command would probably have gained a decisive success* [emphasis added].

Baker's star as an Indian fighter shone out brilliantly on the Marias River in January, 1870, but suffered a great diminution of luster on the Yellowstone in August, 1872. The troops in this engagement suffered a loss of only one killed—Sergeant McLaren, Co. C, 7th Infantry, and three severely wounded—privates O'Malley, Co. E, 7th Infantry, and Ward, Co. L, and Cox, Co. F, 2nd Cavalry. One of the citizens, Francis, was also wounded and died three days after the engagement.

After this affair the troops continued their march slowly down the north bank of the Yellowstone, the engineers carrying forward their survey; but Engineer Haydon's fears had got the better of him, and he sought an occasion to return. He sounded the opinions of the officers, but found the majority of them in favor of pushing on and satisfied of their ability to take care of themselves. At last, on the twentieth of August, at a point about six miles above Pompey's Pillar, he insisted upon returning or turning off toward the Musselshell, and the latter course was pursued. After surveying across the country to that stream and up its south fork, the expedition finally disbanded on the twenty-fifth of September, the troops returning to their posts.

Engineer Haydon, though wholly responsible for the failure to prosecute the survey to Powder River as had been originally designed, afterwards endeavored to shirk it upon the military. Had it been his desire to proceed; there would have been no hesitation on the part of the commander of the troops to accompany him; and the great majority of the officers were eager to go on, to save that command from any suspicion of having been frightened from its purpose by Indian hostility.

Source: Bradley, March of the Montana Column, *56–63. Paragraphs running a full page or more have been separated for ease of reading.*

John W. Ponsford's 1910 Account

John William Ponsford (1847–1912) appears to have been raised in Wisconsin, joined the Second Cavalry at a young age, and possibly become a noncommissioned officer when he participated in the Marias Creek Massacre (1870). Two years later he took part in the western survey, although it is not clear whether he was still in the cavalry. Regardless, after leaving the cavalry he stayed in the Bozeman area, being a frontier Jack-of-all-trades: hunter, scout, prospector, and builder. By the 1880s he was a saloon keeper, served as Bozeman's sheriff for a few years, and near the end of his life owned a liquor store.

This account has all the weaknesses of a "remembrance" dictated thirty-eight years after the incident described. However, there is no question that he participated in the Baker battle, and this account offers an exceptionally dramatic, first-person rendering of the Indian attack and why it failed. Ponsford dictated and signed this narrative on November 15, 1910.

In 1871 four troops of United States cavalry, F, G, H, and L, were detailed to escort the surveyors of the Northern Pacific Railway. They left Port Ellis and went down the Yellowstone until the cold weather drove them back. No hostile Indians were seen on this trip.

In 1872 the same detail of troops were made for the same purpose, with the addition of four companies of the Seventh United States Infantry. The command, when it arrived at Sweet Grass [Creek], went up Sweet Grass and then took an easterly course to keep out of the bottoms and avoid crossing the Yellowstone. The surveyors took up their work from where they finished in '71. On August 12th the command camped about ten miles north of where Billings now is and three men—Joe Hozay, a half-breed guide, and two cavalrymen—went out hunting. About noon they saw a band of buffalo on Alkali Creek and before going after them concluded to rest their horses and eat a lunch. While resting, a horse without any rope or anything on, trotted up to their horses, which were picketed, and ran around them. Joe Hozay lassoed the strange horse and exclaimed, "It is a Sioux horse," which immediately put them on their guard.[67]

[67]Note in original: "Six months afterward, it was learned that three Sioux scouts had been watching them from some brush nearby and had aimed their rifles at them. They did not fire for fear that they might miss and the report of their guns would bring more soldiers who would undoubtedly capture them. These three Indian scouts had been sent out by the Sioux force as spies and had been following the camp right along."

Joe Hozay and the others did not get to camp that night until nearly midnight on account of the pack horse running with the buffalo down the Yellowstone. This Sioux horse incident was the first intimation that there were Sioux in the neighborhood.

The next day the command moved to the Yellowstone River, and camped above the mouth of Pryor Creek on what is now known as Baker's battle field. There were some camp followers following the command—prospectors and wolfers,[68] who made their beds and slept outside of the lines of the guards at night. One of these, Jack Gorman,[69] before going to bed, propped his rifle up against a tree, but his cartridges and revolver were in the blankets with him.

During the night (3 A.M.) he was lying awake and saw what he thought was a bunch of feathers moving and, watching for a second or two, found that the feathers had an Indian's head in them. He cocked his revolver and without moving his body shot the Indian through the head. The report of his pistol awoke the camp and the Indians commenced shooting at the soldiers but shot very high.

The soldiers rushed out of the tents, the cavalry to the picket lines and saddled up. The Indians were shooting in at us and the soldiers were shooting back, and this kind of fighting was kept up until break of day when the Indians retired to the bench land [plateau] one-half mile away. They got away with five head of horses and a small beef herd. Several Indians made dashes down from two points of the bench, daring the soldiers to come out in the open, and kept it up until one of them got shot. Mitch Boyer, who was a guide employed by the United States, was with the command and the Sioux Indians called for him to show himself and [said] they would skin him alive.

About this time a call was made for twenty volunteers to make a rush onto the bench to find out the number of the Indians. They made a dash onto the bench and while there the Indians made a dash from each point of the bench, trying to cut off those twenty men from the main command; but the move was seen by those in camp and they rushed out on foot and drove them back while the skirmishers returned. The twenty men who volunteered reported from 1,200 to 1,500 Indian warriors.

Our casualties were very few. A citizen teamster was killed, Private

---

[68]"Wolfers" were trappers who specialized in wolf pelts.

[69]Unidentified, but because he was wanted, "Gorman" was undoubtedly a pseudonym.

Cox was shot through the stomach, Ward of L Company was shot in the eye. The volleys shot in the dark at the Indians did a great deal of damage, killed plenty of them and wounded more.

About 10 A.M. the cavalry and infantry moved out onto the bench; the Indians moving down the river, taking their dead and wounded with them and for transportation of them made travois and rafts.

We made a blind camp and then about dusk moved a mile or two; dug rifle pits and went into camp for the night; but the Indians did not show up again. No more surveying was done that season and the command returned by way of the Musselshell to Fort Ellis. The cavalry was in command of Col. E. M. Baker; the infantry was in command of Capt. C. C. Rawn.

It was heard afterwards, at Fort Ellis, that these Sioux were on their way to Gallatin valley to clean out the valley but having run across such a large bunch of horses and mules concluded to take them in first. In the councils of the Sioux it was a question with them whether to attack us or let us keep on down the Yellowstone and come to the Gallatin valley while the troops were absent. The surveying party was in charge of Major Mullenberg.[70] The Sioux were composed of several different bands—Sitting Bull's Hunkpapa, Brulés, Oglala, etc. It seemed as if they had moved from all over the country to concentrate for this fight.

The fact that Jack Gorman had placed his rifle up against a tree and an Indian saw it and was crawling up to steal it, and that Gorman was awake at the time, is the accident that probably saved the whole command from being killed. The Indian that Gorman killed was crawling ahead of the big outfit of Indians. He had on just his breech cloth, moccasins, and war bonnet.

Late at night of the 13th dispatches were brought from Fort Ellis to the commanding officer of the expedition, by Ordinance Sergeant O'Neil, and among them was one ordering the arrest of Gorman for horse stealing. He was put under guard on the 16th of August and on the night of the 18th, at a camp on the Musselshell River, escaped from the guard just before dark and went to Fort Benton.[71]

---

[70]An interesting mistake: Edward F. Muhlenberg (d. 1883) was the engineer in charge of the 1871 survey. Because of his drinking, Milnor Roberts never rehired him.

[71]There is a high likelihood Baker allowed Gorman to escape as a reward for his having discovered the Indians and killed one.

Richard Davis, now of Bozeman, was guarding the quartermaster herd and helped drive in the mules, but the Indians got away with the steers and killed them and shook their hides at us during the morning.

The Indians, before the fight and during the darkness, imitated by their calls, ducks, geese, coyotes and other wild animals. They used these signals to keep themselves located. On one side of the camp was the Yellowstone River and on the eastern part there was a slough, brush and cottonwood trees. This slough prevented a quick attack on the soldiers after the first shot and it was there that most of the Indians were killed and wounded.

The Sioux, a year or so afterwards, when asked how many were killed, would not tell. Taut said "a heap." They probably concluded not to make any more night attacks, and took daylight for theirs, as they did when they made the Custer fight.

Source: Burlingame Special Collections Library, Collections #659, Montana State University, Bozeman.

Lt. Gen. Philip H. Sheridan (1831–1888)
As commander of the Division of the Missouri, he was militarily responsible for an area that ran from Mexico to Canada and from the Mississippi through the Rockies, including the Dakota and Montana Territories. Lincoln said Sheridan was so short that his arms touched the ground when he stood. While the photo was taken in 1864, it is possibly the most flattering picture of Sheridan ever taken and fully reflects his intensity and leadership qualities. *Courtesy National Archives.*

Maj. Gen. Winfield Scott Hancock (1824–1886)
Hancock commanded the Department of Dakota (Dakota, Minnesota, and Montana). The popular hero of Gettysburg (for stopping Pickett's Charge and being gravely wounded), he had been a strong candidate to receive the Democratic presidential nomination in 1868. He was not on speaking terms with President Grant (a West Point classmate) and held Sheridan in contempt. He was lukewarm about the Northern Pacific until Gregory Smith began searching for a job for his younger brother. *Author's collection.*

Col. David S. Stanley (1828–1902)

Stanley commanded the District of Dakota. An effective general during the war (reaching corps command), he was a careful but brave commander. In 1848 he and Sheridan traveled to West Point together, forming a special bond. Sheridan chose Stanley as one of his generals when an 1865 invasion of Mexico to oust the French seemed imminent. His inherent caution, knowledge of the Northern Plains, and loyalty to Sheridan made him a solid, safe choice to lead the infantry-only eastern survey's soldiers. *Author's collection.*

Col. John Gibbon (1827–1897),
commander of the District of Montana

A fierce, effective Civil War combat general who was to command a corps, he was badly wounded at Gettysburg, leading the Indians to call him "one who limps." Gibbon was so highly respected that Grant appointed him as one of the three "commissioners" to receive Lee's surrender at Appomattox. Although he was a logical choice to lead the escort for the western survey, Sheridan passed him over in favor of Baker. *Author's collection.*

MAJ. EUGENE M. BAKER (1837–1884), COMMANDER OF FORT ELLIS (FOUR COMPANIES OF CAVALRY AND ONE OF INFANTRY)

Baker was a personal favorite of Sheridan's dating back to the '64 Shenandoah valley campaign. An alcoholic and sadist who led the Marias Creek massacre, he was too inebriated to leave his tent during the Indians' surprise August 14 predawn attack. His duplicity in forcing the western survey to turn back had disastrous consequences for the Northern Pacific. Photograph taken c. 1859. *Courtesy United States Military Academy, West Point, N.Y.*

MAJ. JOHN W. BARLOW (1838–1914)

Barlow was an artillerist and combat engineer during the war. He became Sheridan's chief engineer and a gifted observer and writer. He accompanied the Hayden Expedition into today's Yellowstone Park in 1871. Traveling with Baker's column on a similar assignment in 1872, he wrote the only known detailed account of this disastrous surveying expedition. Sheridan made him rewrite his report (was it too explicit concerning Baker?), but this remained sufficiently damning that Sheridan later sent him to the Corps of Engineers, where he languished until Sheridan's retirement. Photograph taken c. 1860. *Courtesy United States Military Academy, West Point, N.Y.*

JAY COOKE (1821–1905)

Cooke, the "Financier of the Civil War," financed more than a quarter of the Union's war effort through the scandal-free sale of bonds. This seldom seen sketch of Cooke, made during or just after the war, captures his intelligent eyes. After the war's end, Cooke was arguably the nation's second most powerful man. This sketch was apparently made in the 1890s, but in it he appears far younger. *Sketch by Robert M. Vonnoh, 1894, from Ellis P. Oberholtzer,* Jay Cooke: Financier of the Civil War *(Philadelphia: George W. Jacobs and Company, 1907), vol. 2, p. 250 (copy in author's collection).*

Northern Pacific president J. Gregory Smith (1818–1891)
A Republican wartime governor of Vermont and longtime believer in a northern transcontinental railroad, Smith was also president of the financially troubled Vermont Central system. Exceptionally able and hardworking, Smith was also ethically challenged and knew little about business management. Whereas it had taken two companies to build the first transcontinental, Smith thought he could manage Northern Pacific construction as well as his own Vermont Central. Even if he had been totally honest, he had bitten off more than any one man could chew in his era. Photograph taken mid-1860s. *Brady photo collection, courtesy National Archives.*

Northern Pacific chief engineer
W. Milnor Roberts (1810–1881)

The Quaker and Pennsylvania-born Roberts began his surveying career before the first American locomotive was invented. His career included canal, road, bridge, and railroad engineering and construction, to say nothing about harbor and related public engineering works. The photo appears to have been taken just before the Civil War began. *Courtesy Montana State University Library, Bozeman.*

EASTERN SURVEY CHIEF ENGINEER T. L. ROSSER (1836–1910)
Rather than graduate from West Point's famed double-class of 1861, he resigned to join the Confederacy. A large, powerful man, he was fierce in battle and rose to become a two-star major general. Rosser's engineering skills were never strong, but his energy and leadership qualities were remarkable, and he was a perfect choice to lead and motivate the surveyors. Photograph taken c. 1872. *Courtesy Albert H. Small Special Collections Library, University of Virginia, Charlottesville.*

Eastern surveyors with Rosser in white tuxedo

This photograph, likely taken in 1872, shows some of Rosser's surveyors (with their equipment) very much at ease and mugging for the camera. *Courtesy Albert H. Small Special Collections Library, University of Virginia, Charlottesville.*

Sitting Bull (1834–1890)

The unquestioned spiritual leader of the "militant" Northern Plains Indians and a great warrior by any standards, more than anyone he led the opposition to the Northern Pacific Railroad in 1872. Little credit is given to his "military intelligence" network, but he apparently was fully aware of the Northern Pacific's primary engineering goal—having both surveys meet at the Powder River—and blocked this from occurring. As a result, both surveys were considered failures and Northern Pacific bond sales effectively ended. *Courtesy National Archives.*

GALL (CA. 1840–1894)

One of the most charismatic of the Sioux warriors, Gall was a Hunkpapa Teton (or Lakota) Sioux, raised by Sitting Bull's parents. Held in high esteem by his fellow Sioux, Gall survived being bayoneted through his torso in the mid-1860s. He went on to play a key role in the 1872 and 1873 Yellowstone surveys. Gall's ambush and killing of Lewis Dent Adair, Julia Dent Grant's first cousin, was a public relations disaster for the Northen Pacific. At the Little Big Horn, two of his wives and three children were murdered by cavalrymen. *Courtesy National Archives.*

O'Fallon Creek battle site, Sitting Bull's position
Having a slow moving column pass in front of you would have been any attacker's dream, but after the Baker battle, most of Sitting Bull's forces went off hunting. With only two hundred men and poor ammunition made by the Sioux themselves, Sitting Bull could do little to stop Stanley's superior force. To the left near the stream bed is approximately where Gatling guns proved unable to hit mounted Indians. *Author's photo.*

O'Fallon Creek battle site, Stanley's position
The cliffs were 150 to 200 feet high. To make matters worse, the Indian position was to Stanley's southeast, causing his troops to have to look into the morning sun. *Author's photo.*

(above) Indian position, Baker battle
The woods seen here are approximately where Baker camped. While a stream or half-moon lake did exist in 1872, the Yellowstone's changing course has undoubtedly altered the current view. The road in the background is Interstate 94. The Yellowstone River is between the trees and the interstate. *Author's photo.*

(opposite) Soldier position, Baker battle
This photo gives a good feeling for the approximate location of Baker's defensive position in the woods the morning after the uncoordinated Indian attack. Before their retreat, the Indians occupied the ridge line to the right where they quickly used up much of their limited ammunition. *Author's photo.*

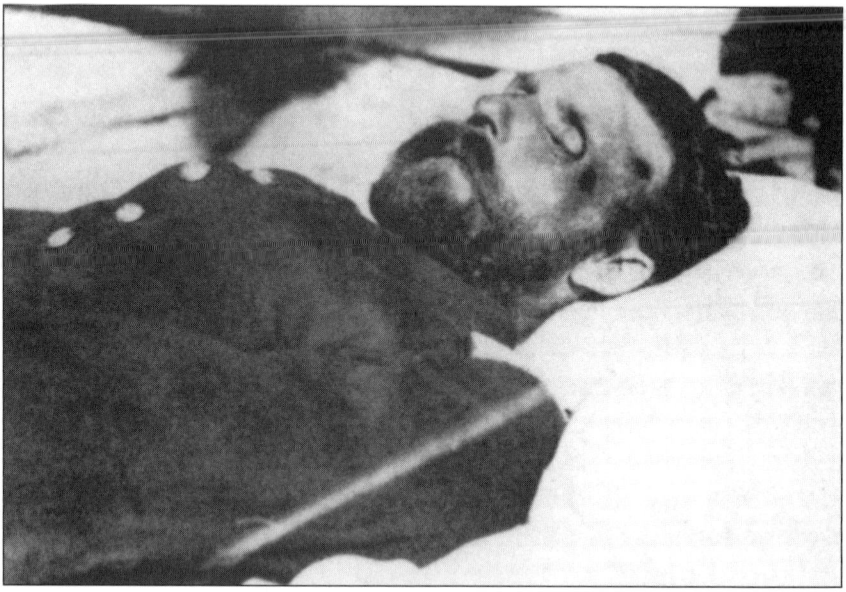

Lt. Lewis Dent Adair's Corpse

This grim and seldom seen photograph is of Lieutenant Adair, the first cousin of President's Grant's wife. The photo was taken at Fort Rice, some days after Adair's death at the hands of Gall's Hunkpapa Indians. Although the body has also been identified as that of Lt. Eban Crosby (killed, scalped, and mutilated a day earlier), the fact that this corpse had not suffered these humiliations points to it being Adair. In addition, it had previously been identified as Adair in 1956. *Author's collection.*

"Death of Lieut. Dent Adair, USA"

Amateur artist and Northern Pacific surveyor Charles "Shorty" Graham (1852–1911) left the railroad in late 1872 and became a lead illustrator for *Harper's Weekly*. This drawing portrays Rosser, the mortally wounded Adair, and the Indian that Rosser was about to shoot in the face. As the sketch appears to accurately reflect the details of the incident as well as the countryside near the Heart River, it was likely drawn within a few days of the attack. *Courtesy Albert H. Small Special Collections Library, University of Virginia, Charlottesville.*

(above) GRAHAM'S DRAWING OF THE HEART RIVER CANYON
Because of the expense of building through the canyon, to say nothing of the extra miles involved, Rosser found another route along Sweet Briar Creek. *Courtesy Albert H. Small Special Collections Library, University of Virginia, Charlottesville.*

(opposite) GRAHAM'S DRAWING OF HEART BUTTE AND HEART BUTTE TODAY
Geographically unique and only a few miles from the Heart River, the butte was a well-known Indian trading and meeting area. *Courtesy Albert H. Small Special Collections Library, University of Virginia, Charlottesville; photo courtesy of author.*

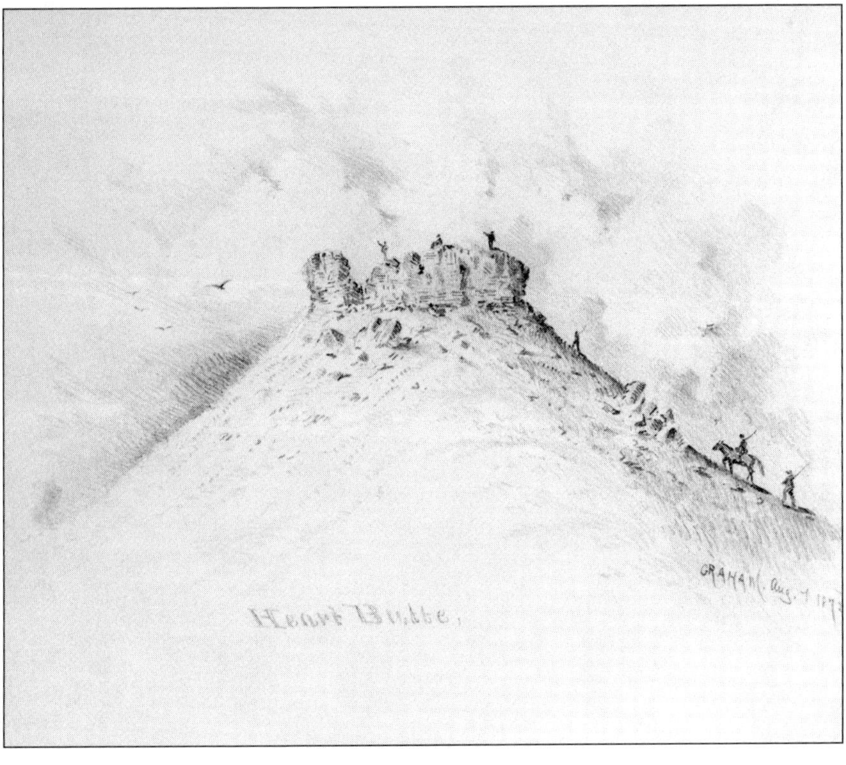
GRAHAM. Aug. 7 187
Heart Butte.

Brainerd depot

This totally unnecessary building, constructed in 1872 as Northern Pacific ran out of cash, replaced a two-story wooden structure barely a year old. *Courtesy Minnesota Historic Society, St. Paul.*

PART III

Bitter Returns

August–November 1872

On August 23 Sitting Bull and Gall ceased harassing the eastern survey. With Stanley having brushed the Indians aside without casualties, the surveyors' attention returned to engineering, thankful that Linsley was returning east. The next evening, just before midnight, three dozen empty wagons, Linsley, Captain Kellogg, Henry Davis (who could no longer walk), and a two-company escort—perhaps 150 men—headed for Fort Rice. The Stanley-Rosser column was now in eastern Montana, twenty miles northwest of today's Baker.[1] In this part of Montana, Rosser would spend three weeks unsuccessfully searching for a Bismarck-to-the-Yellowstone shortcut in almost desertlike country. While we now know that no such route existed, from reading Rosser's diary and surveyor comments one presumes Rosser felt that he would be discharged if he failed; his increasingly frantic actions were consistent with a man fearful for his job.

In contrast to Baker's relationship with Haydon, Stanley, if reluctant, fully realized that his role included full support of the surveyors, and he therefore acquiesced to Rosser's engineering requests. With alkaline water, heavily salted rations, little fresh meat, few trees, and poor forage, men and animals quickly gave out. Making life even more difficult for infantrymen, their shoes were shoddily manufactured, and the abundant small cactus added to their unhappiness. As Nibsey Swipes commented in the *Sioux City Daily Journal*, the men were beginning to look like Falstaff's ragamuffins.

* * *

Once at Pompey's Pillar the western column cut north across some fifty miles of treeless land, crossing the Yellowstone-Musselshell divide on August 22. In the Musselshell Valley trees began appearing again and the next day they reached the river itself. The river, named for its abundance of fossilized shells, is a placid stream flowing from the Crazy Mountains, a hundred miles west, but its water is laced with diuretic chemicals. Working their way up the valley, which was devoid of natural obstacles and filled with game and buffalo, it took the disheartened column over five weeks, barely five daily miles, to return.

---

[1]Baker, Montana, is named for a railroad executive with the Chicago, Milwaukee, St. Paul & Pacific in the early 1900s and not Col. E. M. Baker.

Initially another attack was feared, but few Indians were seen north of the Yellowstone. Just before the survey ended, their slow pace combined with increasingly higher altitudes saw them engulfed by an early snowstorm some thirty miles south of White Sulfur Springs.

* * *

Finally Rosser gave up on finding a shortcut. After crossing the Little Missouri and making good time, he suggested to Stanley that, with the weather becoming colder, why not release three Eighth Infantry companies who had been issued particularly shabby tents. Stanley agreed and on October 1 three companies under Major Crofton headed east.

Crofton was twenty miles ahead of Stanley when he found a good campsite on October 3. Shortly afterward the popular, one-armed Lt. Eban Crosby[2] went hunting alone as perhaps a fifty-man war party under Gall silently watched. Crosby was followed and, once out of Crofton's sight, was captured and likely tortured.[3] No one was even aware that he was missing until Gall galloped close to the camp, waving Crosby's fresh scalp on the end of a spear or coup stick. Search parties were immediately sent out, but didn't find his body until the next day. Strangely, Crofton never sent a messenger to Stanley with the news, which, with twenty-twenty hindsight, was a shocking oversight.

Stanley's column marched east, ignorant of Gall's attack. No Indians had been seen in a month, and with Fort Rice less than sixty miles away, the column's discipline had become lax. Rosser was scouting by himself, northeast of the column, when Gall's warriors suddenly appeared, racing *between* him and Stanley. Mounted on a draft horse, he kept calm, pointing his rifle at any Indian who came too close, firing occasionally but retreating *away* from Stanley. Gall then sent a half dozen warriors around a butte to block Rosser, only to come upon Lt. Lewis Dent Adair, the first cousin of Julia Dent Grant, wife of the president. Three Indians dismounted and, in the first exchange of gunfire, the Indians missed

---

[2]First Lt. Eban Crosby (ca. 1840–72) had lost a finger at Gettysburg. The wound was poorly treated, and eventually his forearm was amputated. From Maine, he enlisted in 1861, gradually advancing in rank and becoming an officer in 1866.

[3]Stanley Vestal, who had access to a handful of survivors, wrote in *Sitting Bull: Champion of the Sioux* (1932, 131), "It is said that Gall collected his remaining arm and took it home as a trophy."

while Adair hit one in the torso. The Indians reloaded first, or perhaps had repeating rifles, but fired before Adair could get off a second shot, hitting him in the spleen. With Rosser's arrival the two unwounded Indians quickly fell back, while Rosser, aiming carefully, shot the wounded Indian in the face. (See figure 19 for an accurate depiction of the incident.)[4]

One unanswered mystery is exactly how long Gall had been following Stanley's column. It is this writer's supposition that Gall's appearance was entirely by accident and that Gall was heading toward Fort McKean when "targets of opportunity" suddenly materialized—far too much of a temptation to pass up. Had Stanley and Rosser not spent over three weeks looking for the Yellowstone shortcut, they would have arrived back on the Missouri some weeks earlier. After Gall's ambushes, he continued on his (likely) original goal: attacking Arickera scouts at Fort McKean, which the surveyors later witnessed.[5]

This section includes Rosser's detailed account written a week later, his "Death of Adair," and other surveyor and newspaper accounts. Rosser later acquired a sketch by surveyor Charles "Shorty" Graham (1852–1911). Graham, self-taught, became a featured *Harper's Weekly* artist and likely drew Adair's death within days. If it is not of high artistic merit, Graham's work accurately portrays the landscape, clothing, and locations of the people involved.

[4]Lewis Dent Adair (ca. 1839–72) had a splendid war record, including leading an Ohio regiment in battle after higher officers were killed or wounded. He joined the Regular Army in 1866. Other than being married with three children, little of a personal nature is known.

[5]Col. Henry Boyd McKean (not McKeen as it is sometimes spelled) was killed at the June 1864 battle of Cold Harbor. He was a Princeton graduate and a friend of Gen. Winfield Scott Hancock, who named the post in honor of McKean's bravery. With Hancock's transfer in late 1872, Sheridan soon changed the post's name to Fort Abraham Lincoln.

The Eastern Survey Returns

Northern Pacific Surveyors' Field Diary,
August 23–October 26

FRI. AUGUST 23RD Clear & very hot—Ran seven miles of line to-day—both parties were out and lapped over each other—Worked fourteen hours—Nothing seen of the Indians to-day.

SAT. AUG. 24TH Clear & hot. Had to go back two miles on yesterday's line and change it. Ran about three miles. Stopped at 1 o'clock and went into camp—Had to unload the teams that are going for supplies. Train for Missouri River started at 11½ P.M. Thirty-seven wagons with an escort of 100 men. Mr. Linsley went in with them.

SUN. AUG. 25TH Cloudy & windy—Both parties laid in camp all day—Gen Rosser was off all the forenoon exploring a creek that runs off from O'Fallon's Creek S.E. An Indian scout from Ft Rice arrived in the afternoon with dispatches for Gen Stanley. He brought word that Baker had had a fight with the Indians at Pryor's Creek on the 14th and been badly whipped although he had succeeded in repulsing the Indians—the scout who brought the dispatches had been only four days on the trip. He reports having seen a few hostile Indians in the Bad Lands.

MON. AUG. 26TH Cloudy—Ran seven miles of line and reached the point where we leave O'Fallon's creek—Eckelson & myself moved into Gen Rosser's tent to-day—We are to mess [eat] with him from this time.

TUES. AUG. 27TH Clear. I did not go out on the line to-day as hereafter Eckelson and myself go out on alternate days—Party ran 8½ miles; camped opposite our camping ground of August 14th.

WED. AUG. 28TH Clear. Gen Rosser with Jordan & party ran back through a new route to O'Fallon's Creek and found it two miles shorter than the other one—I ran four miles up a creek opposite our camp—trying to find a break in the bluffs to get through but did not succeed—Four Indians reported to have been seen—Train did not move.

THURS. AUG. 29TH Clear & warm—Capt. Goodloe with his company went ahead of the train to the place where we camped to dig wells as there is a great scarcity of water—The water did not hold out to water all the stock.

FRI. AUG. 30TH Clear & very warm—Train did not move today—Party ran about five miles—Line will probably have to be abandoned as the ground rises too rapidly to get a grade line up.

SAT. AUG. 31ST Clear & warm—Reached Cabin Creek about two o'clock—Indians reported today & train corralled—Line party quit about three miles from camp.

SUN. SEPT. 1ST Clear—Laid in camp all day. Gen. Rosser went out exploring with the scouts.

MON. SEPT. 2ND Rained very hard indeed from two to four in the morning and raised the creek very high—The traveling was very muddy indeed and made it very hard running. Line was turned down Cabin creek towards the Yellowstone as we have got to abandon the line we have been running—Ran twelve miles.

TUES. SEPT. 3RD Clear and intensely hot had some very hard pulls and a great many crossings of the creek to make—Line party ran twelve miles.

WED. SEPT. 4TH Clear & intensely hot—Ran 8½ miles; the last part of the way was through very rough country and will not do for line.

THURS. SEPT. 5TH Had a tremendous rain & hail storm just after breaking camp and it was showery through the day—The line party went up to the Indian grave just west of O'Fallon's Creek and connected with the line there and ran east.

FRI. SEPT. 6TH Cloudy, windy and cool—I went out with the party to-day and we had very good luck with the line. Ran up on to the plateau—Found a quantity of plums on the creek up which the line was running and gave the creek the name of Wild Plum. Had an alarm of Indians in the forenoon but it proved to be a false one.

SAT. SEPT. 7TH Clear—Stayed with the train—Went into camp at noon on Sandy Creek—Line party connected with our line down Cabin Creek and got good grades.

SUN. SEPT. 8TH Clear—Marched twenty two miles—Line party did not Work—One of Gallagher's leaders [mules] broke his wiffle tree [part of the harness system for wagons] coming up a hill and got away and charged at everything visible. He knocked one man down and injured him slightly.

[*After this date, finding a good route became unlikely. The bad news is shown in italics.*]

MON. SEPT. 9TH Cold & showery—Had *poor luck with the line in the forenoon and had to abandon it*—Gen Rosser was out exploring the Bad Lands for four hours.

TUES. SEPT. 10TH Clear—Train went into camp at noon—Had fine grazing and good water—The line party worked all day and got into the Bad Lands—*It begins to look very dubious for our prospects of getting the line through near here.*

WED. SEPT. 11TH Clear—Camp did not move—*Had to abandon the line again* as we got into an almost impassable country—the roughest and most broken country that we have struck yet.

THURS. SEPT. 12TH Clear—The line party took tip the line on Cabin Creek and ran up the creek 13½ miles and found a good line as far as they went—Train moved at five o'clock and camped at eleven—Fair grazing and good water.

FRI. SEPT. 13TH Clear & warm—Backed up on yesterday's line three miles—staid with the party until noon when the General sent me in to connect the A line with the L line—Had to walk six miles to do it—*The line party got into rough country again; it seems as if we were never to get out of this cursed creek.*

SAT. SEPT. 14TH Clear—The line party ran until past noon and reached the top of the bluffs where the line was abandoned as impractical. The view of the Bad Lands from the top of the hills was very fine indeed—The train travelled until five o'clock and camped on Truman's Creek.

SUN. SEPT. 15TH Clear & cool. Whole outfit laid still except Gen. Rosser who took the scouts and was gone from 5½ in the morning until seven in the evening exploring the country—They rode at least sixty miles but did not find any route for the line. The Gen. has decided to run a line over the Bad lands and have a tunnel at the summit.

MON. SEPT. 16TH Clear—Started a line about two miles west of Truman's creek and ran 4½ miles, *the last mile through terrible rough country in the Bad Lands—Made the survey with the view of getting a location for a tunnel as it has been found impossible to get a grade line over the summit unless the* [*very expensive*] *tunnel plan is adopted.*

TUES. SEPT. 17TH Clear and windy—Camp did not move—set up stove in tent for the first time. *Line party tried another route to-day but it was not successful.*

WED. SEPT. 18TH Clear & pleasant—Started a new line from the summit and ran 200 Sta *through a very rough country where we were able to get a grade line but the work* [*would be*] *tremendous.* Camp did not move.

THURS. SEPT. 19TH Snow squall in the morning but clear by 8 o'clock—Clear—Line party continued the line of yesterday and connected it with our survey up Cabin Creek and then returned to the summit and ran east—Struck tents at 8½ P.M. and moved four miles easterly down the creek.

FRI. SEPT. 20TH Clear—Ran nine miles of line—Camp did not move—I did not go out today as I was not feeling well—Scouts reported seeing two horsemen in the direction of the Little Missouri but were not able to get within speaking distance as they ran from them. They are supposed to be scouts from Reed's party as they were riding shod horses.

SAT. SEPT. 21ST Clear—Ran nine miles and stopped two miles east of Pettit creek. Eckelson came out with some mail—Reed's train is at the Little Missouri and Davis brought our mail over—Gen Rosser received news of the death of his youngest child.

SUN. SEPT. 22ND Clear & warm—Train moved to the Little Missouri and joined Reed's party—It was pretty lively around camp [from all the drinking] in the evening.

MON. SEPT. 23RD Rainy—No work done on the line to-day—Took account of freight that came out and loaded the teams that are to stay.

Good many men were full to-day—Donha got so wild and abusive that Eckelson had to send him to the guard house—*Gen S got full and Eck and myself gave up our bed to him.*

TUES. SEPT. 24TH Rainy and very cold wind—Ran five miles; last three in the valley of the Little Missouri.

WED. SEPT. 25TH Clear but very windy and disagreeable—Camp did not move as they were busy all day fitting out the train to go in to Fort Rice to-morrow—Line party went back and altered the line; threw out two river crossings by greatly increasing the amount of curvature.

THURS. SEPT. 26TH Clear & pleasant—Train for Fort Rice left at daybreak—Donha has been discharged and went in with it. Taylor & Davis also went in both leaving of their own will—Ran four miles of line to-day in Little Missouri valley and made five river crossings. Had two miles of heavy traveling to make camp.

FRI. SEPT. 27TH Cloudy & warm—Moved camp about two miles—Line party left the valley of the L. Missouri to-day & ran two miles up the creek on which we were camped—Eastman got hold of some whisky today and got full—

SAT. SEPT. 28TH Cold windy day with frequent heavy showers—Ran four miles & quit work at 3½—Got a very good grade line but were obliged to exceed the limit of curvature in one place where we put in a 4° curve—Camped on the creek where Linsley and I had the row coming out.

SUN. SEPT. 29TH Clear & warm—Line party worked to-day as Gen. Rosser is anxious to get out of the Bad Lands and they passed the summit to-day making a nine mile run.

MON. SEPT. 30TH Clear & warm—Only a small portion of the line party worked and they ran only two miles to bring the line up opposite to camp—Gen. Rosser went off exploring with the scouts and was gone all day. Eckelson, Major Worth & myself went off hunting antelope in the afternoon and rode about twenty miles but [did] not kill anything.

TUES. OCT. 1ST Clear & very warm. Jordan had to stay in camp to do some work so I ran his instrument—We started at 7 o'clock and by 11½ had 8 miles run, quit work at 4½ after a run of 13½ miles, the longest run yet made.

WED. OCT. 2ND Clear—Line party ran ten miles—Eckelson ran Jordan's instrument—

THURS. OCT. 3RD Clear & pleasant until 3 o'clock when the wind shifted to the N.W. & blew very hard—I ran the transit again—Ran thirteen miles.

FRI. OCT. 4TH Clear & warm. Camp moved nineteen miles to find wood—Went into camp at two o'clock—About three a scout was chased into camp by Indians and a small party of them were plainly seen from camp—Gen. Rosser had a very narrow escape as he was two miles ahead of the line party when he discovered the Indians who numbered about 100. They did not rush on him as they evidently thought they had a sure thing of it—Lieut. Adair who was out with the escort that day was a mile away from the command on foot hunting. Gen. Rosser saw him and motioned him to run. Eckelson joined the Gen. just then and they turned back to help Adair whom they had not recognized but thought was a private soldier. When they reached him they found he had been wounded and had wounded one of the Indians. There were three Indians within five hundred feet of him when Rosser got there and he killed the one that Adair had wounded and got all his trappings off of him—The fight occurred six miles from camp—The ambulance and a company of soldiers were sent out to bring Adair in. Steve (Gen. Stanley's mulatto cook) was caught by the Indians while out hunting and when found had three arrows in his body and five gunshot wounds in his head and neck.

SAT. OCT. 5TH Clear—Adair died this morning at 4:40—Want back to where the fight occurred to start the line. Ran seven miles did not seen anything of the Indians.

SUN. OCT 6TH Clear—Camp moved to the east side of Heart river—Gen Stanley's cook was buried at 4 o'clock—Line party ran the line to Heart river & stopped at 5 o'clock.

MON. OCT. 7TH Clear & warm—Ran the line six miles down Heart River—The pickets reported in camp that we were attacked on line and the scouts started out to our assistance. Did not reach camp until half an hour after dark—Two of the scouts reported seeing fresh tracks of about 100 Indians near Heart Butte going east.

TUES. OCT. 8TH Clear & windy—Line party ran eight miles—

WED. OCT. 9TH Clear, cold & very windy and disagreeable—Reed staid in camp and I ran his transit. Made only about four miles on line. Lazelle's [supply] train arrived at noon. They report that Lieut. Crosby of Crofton's battalion was killed and scalped by the Indians on Heart River on last Thursday.

THURS. OCT. 10TH Clear & warm—Camp moved to the first crossing of Heart River—Line party ran 7¼ miles—Louis Agard left for Ft Rice at sunset with the mail.

FRI. OCT. 11TH Clear & warm—Ran 5¾ miles.

SAT. OCT. 12TH Clear & warm—camp moved to the Big Muddy—Capt. Pearson's[6] Company from Fort Rice under command of Lt. O'Neil joined us to-day—They are to act as our escort after Gen. Stanley leaves.

SUN. OCT. 13TH Clear & warm—Eckelson & self hunted prairie chicken in the morning got six—Gen Rosser decided to start for the crossing in the morning escorted by the scouts—Passed the evening in Worth's tent—

MON. OCT. 14 Clear & warm, a perfect Indian summer day—Gen Stanley and his train left us this morning and we are left with Lt O'Brien and his company—Ran 8½ miles and quit at 3:20 as we could not see camp and were afraid that we might have trouble in finding it—I left Jordan in camp and I ran his transit all day—

TUES. OCT. 15 Cloudy & cool—Ran about seven miles—Saw a large Indian trail and followed it all through the day—Also saw where they had camped within two days—In the evening while Eckelson and I were talking together we heard a horseman ride into camp and upon going out found it was Lieut. Webster—He reported that there had been fighting at Fort McKean yesterday and *Gen Rosser had sent to Gen Stanley asking him to send out a reinforcement of fifteen men and he sent out*

[6]The Big Muddy is a creek flowing into the Heart River. Captain and civil engineer Edward P. Pearson (1837–1915) enlisted in 1861. He survived a dozen major battles in the eastern and western theaters, also participating in Sherman's March to the Sea. He was on the 1873 survey, appointed major shortly after, and in 1891, the colonel of a regiment. He married the much younger daughter of another officer. Pearson fought at San Juan Hill but shortly thereafter was stricken with malaria and resigned.

severity-five under Capt. Miner, Conway, & Webster.[7] They brought 14 wagons and the big Gatling gun.

WED. OCT. 16TH Clear Ran seven miles of line, had considerable brush to go through in the first part of the day.

THURS. OCT. 17TH Clear & warm—Genl. Rosser came out to our camp about noon with supplies for Eckelson & men to relieve Miner. He went back in the afternoon with Miner & his command—Ran seven miles.

FRI. OCT. 18TH Clear & warm—Ran 3½ miles and then had to quit as the fire got into the valley *ahead* of us—Found Capt. Powell of the 6th with twenty-five men at camp to escort me in to the river.

SAT. OCT. 19TH Clear & warm—Separated from Eckelson's party at seven o'clock and started for Fort McKean. Reached McKean at 11½—Pitched camp in the bottom[land]—Lunched with Capt Powell—These military fellows at this post are the worst frightened men on the Indian subject that I have seen.

SUN. OCT 20TH Clear & warm—Intended to have taken up the line again to-day but Smith got drunk and was not on hand in time so I could not—Jordan & myself went over to the east side of the river. Stayed there until eleven—Broke the axle to a wagon on our way back—Got on board the steamer *Key West* to cross the river but she got wind bound and we did not get across for five hours and had to cross finally by the ferry.

MON. OCT. 21ST Clear, windy & cold—Went up to the Ft at six o'clock for our escort and had to wait on the hill until seven o'clock before it was ready—Had a Lieut. & twenty men. Marched nine miles to the line and did not get to work until eleven o'clock—Ran three miles—

TUES. OCT. 22ND Clear & warm—Had a sergeant and ten men for escort—Connected our line with [Montgomery] Meigs in the afternoon—Scout brought up word to Col. Carlin[8] that 250 Indians are coming to attack this post.

---

[7]An interesting look at Stanley's cautious thinking process: even with the survey almost over, he took no chances of something else going wrong.

[8]Lt. Col. William P. Carlin (1829–1903), West Point '50. Commandant of Fort McKean/Fort Abraham Lincoln until replaced by Custer the following year. A brevetted major general during the war, he commanded a division during Sherman's final campaigns. Although he retired as a brigadier general, his career became sidetracked after the war.

Rosser Diary, August 25–October 19

SUNDAY, AUGUST 25. My men rest today whilst I examine a creek that comes in from the east at this point. Took the Indian scouts and 10 infantry mounted on mules and rode about 30 miles. [Indian] Scout[9] from Rice with dispatches from St. Paul reached us about 4 P.M. Baker had a severe fight with the Indians on the 14th instant.

AUGUST 26. Ran about 7 miles up O'Fallon's Creek, got good line out [of] the work, was very tedious. I am suffering dreadfully of blues. Weather mild and sky overcast. Such weather is always depressing to me. No Indians seen today. We think they have gone off after Baker.

AUGUST 27. Produced the line up the Sandstone Creek about 8 miles. The line not satisfactory and will have to return and examine the ground in the angle between the mouth of Sandstone and O'Fallon's Creek. Feeling tolerably well today.

AUGUST 28. Camp is about six miles from the point of beginning of work and I take an escort of our company in wagons and go back and run in a cut off. Found the line good and the country far better than I expected. Found a spring on the hill near a sand patch. Returned to camp 4 P.M. Saw no Indians.

AUGUST 29. Camp moved to Sulfur Spring. I took the cavalry and explored the ridge between Cabin Creek and Sandstone [Creek]. Finally found a crossing but little or no water. Joined the party about 2 P.M. and remained with it until night. Found water and [——] at camp and recommended stock be sent to water tomorrow and camp remain where it is.

AUGUST 30. Camp remained at Sulfur Spring and with party I ran the line to the divide. Found the grades heavy and lines good. Returned to camp and to my surprise found that Eckelson had dug a spring which gave us plenty of water. Saw the teamsters had got plenty of grass and stock had fared well.

---

[9]The scout was Goose, with the army since 1870. Photos of Goose survive, including one with Custer and Bloody Knife in 1874. Some sources refer to him as "Arickera" while others call him—or possibly another man—a "Yanktoni" Sioux with the name Drifting Goose. A scout by the name of Goose also fought with Custer in 1876, but he survived.

AUGUST 31. Ran line over the hill into Cabin Creek. *Found grades heavy and alignment bad.*[10] Very warm and water bad, men and horses suffered. Camp moved to Cabin Creek. Saw a few Indians and had a halt for the expected fight which did not come off.

SUNDAY, SEPTEMBER 1. Seeing no way of getting out of Cabin Creek which yesterday's work brought me into I gave the party rest. And with the Indian scouts went out to look the country over and to see how I should direct future operations. Failed to find an [engineering] escape and returned and requested the Commander to go down the Creek to the Yellowstone. Yes, [but] with much difficulty.

MONDAY, SEPTEMBER 2. Had a very severe rain shower just before reveille this A.M. which raised the creek and set all the ravines to discharging large amounts of water. We began work running down the creek. The mud in the Bad Lands was fearful and work was tedious yet we ran twelve miles.

SEPTEMBER 3. Very warm day. Party suffered very much from heat but the country was good and we ran not quite 12 miles. But owing from the difficulty encountered by the wagon train which encamped early, we did not run an entire day. Line good and we are all hopeful.

SEPTEMBER 4. Set out early this A.M. for the Yellowstone. Came in sight of a creek which at first sight we thought was the Yellowstone but when we reached it we were disappointed. *Ran 4 miles further over very rough country without success and abandoned the line.* Will go to the west end tomorrow.

SEPTEMBER 5. Made connection with line on Yellowstone west of O'Fallon's Creek and ran down into a coulee which we followed nearly out to the table land. Think we will get a line which will be satisfactory upon a more careful examination. Had a violent hail storm 9 A.M. and rain occasionally all day.

SEPTEMBER 6. Completed line up Wild Plum Creek from the Yellowstone Valley. Line good but can be improved upon a careful survey by using 5° curves. Day overcast and occasional shower. Quite cold and windy. Soldiers used Buffalo chips for fuel.

---

[10]Sections discussing problems in finding a shortcut are *italicized.*

SEPTEMBER 7. Finished the cut off line via Wild Plum and Sandy Creeks and made connection on Cabin Creek. Line far better than expected. Not well today. Bright, warm day. Nothing of interest occurred on line or in camp. The week has been most profitably spent in making this return trip to the Yellowstone.

SUNDAY, SEPTEMBER 8. Marched 22 miles and encamped at Kangaroo [Mouse] Creek.

SEPTEMBER 9. Resumed line at the mouth of Kangaroo Creek *but before producing it far I discovered it to be a failure and rode ahead to examine the country which I found to be so rough as to put the running of a line out of the question. I returned and turned off more to the East. Broke my horse down.*

SEPTEMBER 10. Continuing line up through the *mauvaises terre. Prospect a gloomy one but hope to succeed* as all men do who undertake an enterprise. Had to run a portion of the line over twice. Made slow progress, only about 2 miles. Clear and cool.

SEPTEMBER 11. Found the line running into the *Mauvaises Terres* and *made a hard ride to the front to examine the country and found it to be impracticable and about 1 P.M. abandoned it. Am greatly annoyed where to go next, will go up the creek.* Clear and pleasant day.

SEPTEMBER 12. Returned to Cabin Creek and resumed line south east. Ran 15½ miles and found but will require a few slight alterations. Walked four miles back to camp which did not reach to dark. *Party very tired but no one more than myself. As my horse is broken down and to ride him is a great labor.*

SEPTEMBER 13. Continued line up the creek. Soon after dinner I rode to the front to see how the country looked. Found it very rough and that I was now high enough. *Ran an angle for grades, which* [*has*] *only* [*a*] *tolerable* [*chance of success*]. Will begin a preliminary [line] tomorrow which I will run to the Little Missouri.

SEPTEMBER 14. Produced the line out of the valley of Cabin Creek. Found it difficult to ascend the hills. *Then had to descend again and cross the Dead Dog Creek out of whose valley I could not get a practicable line and I suspended operations in that direction* and moved over to Inman's Fork [Beaver Creek].

SUNDAY, SEPTEMBER 15. Took the Indian scouts and rode down Inman's Fork and across to Cedar [Canyon] thence north to western view of the Yellowstone. Saw no way of getting a line north and after a hard day's ride from 5 A.M. to 5 P.M. without leaving the saddle I returned to camp quite exhausted. Will now try a tunnel through the hill.[11]

SEPTEMBER 16. Began making examination for a tunnel and decided to make it through Dog's Neck. Ran line on each side but not exactly in the right place and will run it over tomorrow. Camp in a good place and quite convenient to line. Day pleasant. Saw one black tail deer.

SEPTEMBER 17. Tried another line down Cedar Canyon, but *found the ground breaking off too fast and after occupying the entire day had to abandon it. Matters look gloomy enough indeed.* Hope I may get a tolerable line yet. *If I fail there will be trouble somewhere.*

SEPTEMBER 18. Abandoned the line of yesterday and Monday and began a new one a little north. Ran four miles and although the profile is very rough the alignment and grade are good. The work very heavy, yet in consequence we may adopt it. Beautiful day. *Am very tired.*

SEPTEMBER 19. Finished the line running down Cedar Canyon and made connection with the "L" line. Very rough but am satisfied that we have got the best line the country affords. Don't think the company will ever build it, but [will] go back to the [1871] Glendive Creek route. Very, very tired. Day clear.

SEPTEMBER 20. Very cold, overcast, and snowing and raining a little. Took the party back to the hill near the tunnel and reran the line down to Inman's Fork and produced it 3½ miles east. Country looking well and I hope to reach the Little Missouri in 3 days. Quit work at 3 P.M.

SEPTEMBER 21. Just before sunset and after a hard day's work I received the heart rendering news of the death of my precious little boy, John Pelham. It is hard to give him up but these griefs are inevitable and we must be prepared to meet them. May God spare the remainder of my family to me.[12]

---

[11]Rosser meant that he had to consider tunneling as an engineering option. Nevertheless, given tunneling's high costs, Rosser's comments indicate how desperate he was.

[12]It was not to happen. Twelve years later, another son ate a spoiled cream pie at a summer picnic and soon died.

SUNDAY, SEPTEMBER 22 Worked hard all day trying to reach the h[——] with the line. Suffering very much with headache. Could not rest last night. How I [——] my dear wife and want so much to be with her. She misses our dear little Pelham more than I do. God bless her.

SEPTEMBER 23. Did not work today. Remained in camp to take an inventory of the stock which [——] up. It is a cold, rainy and disagreeable day. This has been the most disagreeable trip I have ever made and I don't think I will make another. I must stay with my family.

SEPTEMBER 24. Resumed working this A.M. Still raining but about 10 A.M. the clouds passed off and the remainder of the day has been bright and cold. Reached the Little Missouri River and saw the line several miles down its valley. *Don't like this route at all and don't much expect to see it built.*

SEPTEMBER 25. A very cold, windy, disagreeable day. Tried to get our escort to move camp but the Commanding Officer was too drunk. Went back and made some alterations in the line. Went into camp at noon and wrote to my precious Lizzy and to D.C.L. Have map prepared for D.C.L. Train goes in tomorrow.

SEPTEMBER 26. Train got off early. Resumed line near camp and worked along down the river as best I could. Axe men [with us] are bad and there was considerable chopping so we got along slowly. Deeply grieved that I am not able to return to my precious and distressed wife, but circumstances won't allow it.

SEPTEMBER 27. Finished the line on the Little Missouri. Made five additional crossings, making eleven crossings in all. *Line rough and to my mind very objectionable.* Overcast and rained a little. Saw good deal of game signs, but saw none. Soldiers killed one deer and one antelope. Soldier died last night.

SEPTEMBER 28. A very cold, rainy disagreeable day. Locating on the Red Butte Creek. Command moved about 3½ miles. Country rough but got a fair line. Found a rich coal mine. Camp quite confined but water and grass good. One soldier came in for a wagon to bring in eight antelope which he had killed. I took a good bath.

SUNDAY, SEPTEMBER 29th Resumed location up the Red Butte Creek. Found the country near the head of it much better than I had expected and got a good line but the grade is a little too steep. The line is so much better than was expected that it may be called good. Camp on the summit tonight.

SEPTEMBER 30. Camp was moved about one mile and I had the party run as far as camp. And I with the scouts rode about 20 miles further east to look over the country. Recommend[ed to] the Commanding Officer to send in a part of the escort. [Frank] Eastman has been drunk for several days. Tonight I offered him a pledge, [but] he declines and I discharge him.

OCTOBER 1. Passed through a level country today lying along the valley of the Cannon Ball [River]. The day was quite warm but the party ran 13 [miles], which is a large day's work. Part of the escort goes in tomorrow and I am up late tonight writing to dear Lizzy and D.C.L. Two men not in from hunting today, fear they are lost.

OCTOBER 2. Found the country very rough today. Endeavored to cross the divide between Heart and Cannon Ball Rivers to get upon the waters of the former but failed and had to rerun about four miles. Made eleven miles however as it was and got into camp about 5 P.M. Day warm.

OCTOBER 3. Ran about thirteen miles today. Country good in the morning but in the afternoon we came upon some rough ground. Killed two antelope. Began raining about 3 P.M. and blowing fearfully so as to make it impossible to keep instruments in adjustment and about 4 P.M. had to quit. Camped on a high hill and very disagreeable night.

OCTOBER 4TH Resumed line near camp and found country somewhat improved. Saw Heart Butte in the A.M. and ran [line] for it. After [lunch] I went ahead to examine the line and ran onto a party of about 50 hostile Indians and came near losing my scalp. Killed one. Lieut. Adair and a Negro killed.[13] A narrow escape for me.[14]

---

[13]Stephen Harris was Stanley's servant. Rosser's comments fully reflect his indifference (at best) toward African Americans. Like so many Southerners, he never questioned slavery and was still blaming Northern abolitionists for causing the war many years later.

[14]Did Rosser's laconic comment reflect the likelihood that his wife would read this later?

OCTOBER 5. Returned to the scene of our Indian skirmish yesterday and resumed the line. The Indians had carried off their dead and had cleared off. And saw nothing of them today. I killed one antelope and badly wounded another. Ran only seven miles. The country rough.

DISTANT FIRES in the west. Heart River in sight.

SUNDAY, OCTOBER 6. Camp had to move to Heart River and rather than come back over the ground we ran the line in. Found the distance greater than we expected and the work took all day. Ran nearly ten miles. Made connection on Heart River with the survey of the last year. Saw a great many antelope and the camp of the Indians who attacked us a day or two ago.

MONDAY, OCTOBER 7. Began line down Heart River. Found the stream very crooked and many crossings necessary. One of the axemen was sick and progress was slow. Saw only one deer. Shot several times at antelope but killed nothing. Dry, clear and very warm. Ran only 6¼ miles. Got in at dark.

OCTOBER 8. Had to work 3 miles up the river to resume line. The [river] being very crooked and a portion of the way heavily timbered, we go on slowly. Besides this the day was windy and cold and it was difficult to set up instruments. Camp moved about eight miles and we had to quit work early and walk to camp.

OCTOBER 9TH Walked back four miles to work. Had Capt. Miner's company for guard.[15] The Capt. was with us on the 4th when we were attacked and is a very indifferent officer. Very cold and windy. Ice in the river all day. We ran only four miles. Supplies came in from Fort Rice. No mail. Telegrams very unsatisfactory and annoying.

OCTOBER 10. This morning is cold but calm and clear. The river is better than yesterday and party made seven miles. Scout went into Rice and left the party about 4 P.M. And went to camp and made a report

[15]Charles Wright Miner (1840–1928) was an enlistee who rose to captain during the war but in an undistinguished Ohio regiment that suffered four times as many deaths from disease as from combat. After the Little Big Horn, he participated in two actions against Sitting Bull and acquitted himself well, being cited for bravery. Miner fought in the Spanish-American War at San Juan Hill and later in the Philippines, retiring as a brigadier general.

of progress to D.C.L. [see "Rosser to Daniel Linsley," following], and wrote a hasty note to dear Lizzy.

OCTOBER 11. The weather continues good. Had some bad ground on the start which delayed us a little but by hard work we were able to make 5¾ miles. Walked two miles back to camp. The old man [Stanley] on a spree and I could not see him.

OCTOBER 12. We are running today on best portion of Heart River near the mouth of Mud Creek. The valley is quite broad and the river makes [a] few cuts into the bluffs. We ran something over seven miles. A beautiful day but saw but few deer, [and] great many chickens. Joined by [replacement infantry] company from Rice.

SUNDAY, OCTOBER 13. Laid in camp all day fixing up maps and correcting and reducing notes. Made General Stanley a map of route and made arrangements to send Eckelson back on the survey of the North Fork of Heart River. I will go into the crossing [Bismarck] tomorrow.

OCTOBER 14. Left camp before daybreak and by cutting across the country reached the crossing about 1 P.M. Found a little skirmish going on at Fort McKean, several scalps were given and taken by the Rees, no danger to the fort. Telegraphed to my precious. All well.

OCTOBER 15. In Edwinton [Bismarck] making arrangements to reinforce the party and rejoin them on Thursday. Can't hear anything from [Northern Pacific corporate headquarters in] N.Y. Things look badly in the Division and I will return to Fargo as soon as possible.[16] Received several sweet messages from Lizzy.

OCTOBER 16. At Edwinton preparing to return to the party on Heart River. Can't hear from New York. Got everything aboard the boat and went aboard myself so as to get an early start tomorrow. It is indeed hard to be so near my sweet Lizzie and not to see her.

OCTOBER 17. The boat did not leave the east shore until sunrise. Then got aground and it was 8 o'clock before we left Fort McKean. Reached the party 12 noon and found the escort had no instructions to remain and had to return for another. Gave orders to Eckelson and will go home tomorrow.

[16]"Things look badly" refers to the both the railroad and Jay Cooke's banking house.

OCTOBER 18. Left Edwinton at Sunrise and drove to Doleson camp for dinner. Rested my horses two hours and then drove to the end of the track where I found the train waiting for me. Left team at the end of the track and ran through to Fargo. Lovely night.

OCTOBER 19TH Reached home [Brainerd] this A.M. at 4 o'clock. Found dear Lizzy waiting for me. All well and glad to see me. Will not work today but spend it with my precious family. Will not resume work before Monday of next week.

Rosser to Daniel Linsley, October 11

The initial phase of this letter is a standard engineering report by Rosser to Daniel C. Linsley (D.C.L.). Suddenly, however, on the fourth page, Rosser begins describing his ambush by a war party led by Gall. Making this letter all the more unusual, Rosser never liked Linsley. Ironically, there is a high likelihood that Linsley, who was not in Minnesota and resigned later in the month, never saw Rosser's letter. Some spelling and grammatical changes have been made, technical details eliminated, and paragraphs have been made where none existed. All underlying is Rosser's.

Heart River, Dakota, October 11, 1872

Dear Sir:

We reached Heart River with our survey on the 6th inst. and made connection with the survey of last year. . . . The line from "Red Butte" to Heart River is very good, but I find the work along Heart River very tedious: the difficulty consists not only in trying to avoid the river, but in so fitting the curves as to avoid the high points around which the river so abruptly bends as to require sharp curves to fit it. The valley of the Cannon Ball is far better than that of the Heart River, but it is too far south, at this point it is at least 12 miles south of us, and Major Lazelle . . . says it is an excellent valley far superior to this, but from the information I gather from him, I think it would be difficult to get from our line to it; there is a high ridge all the way from its mouth to the point where we leave it near "Rainy Butte" which it would be very difficult to pass with a railroad.

If the Glendive Creek route should be adopted, I would most earnestly recommend that a line should be run up the North Fork of Heart River. . . .[17]

There are no mountain ranges in this country that I have seen and the only broken [——] that we meet are near streams and have been cut away by erosion, and when we get away from water, we find a level country. I would have examined the valley of the Cannon Ball more extensively and also the creek running into the [Little] Missouri near the mouth of Anderson creek, if I could have been furnished a sufficient escort. My reconnaissances on Cabin Creek and the Little Missouri River completely used up the horses of the scouts and now there are not half a dozen of them mounted.

Since the scout horses broke down I have done my exploring alone and on [October 4] came near making my last reconnaissance for life. The country was considerably chopped up and I had been ahead all day looking up the line, and about 4 P.M. I was riding alone about three miles in advance of the party and its escort, when a party of about one hundred Indian rushed out from behind a hill, their ponies at full speed, their lances flourishing and came at me with all the fury in their savage nature. When I first saw them, they were not over a thousand feet away—if you will indulge me I will attempt to picture the situation; but to give it the life in which I saw it is beyond human expression, and sincerely hope that you may never witness a scene so akin to it, as to enable you to appreciate my situation.

My eyes swept over the prairie in vain for friendly support, none was to be seen; a few antelope stood on the hill close by apparently amazed at this strange conduct of human creatures! I was poorly mounted and knew it would be useless to run, as their ponies would soon overtake me, and there appeared but little possibility of escape from a horrible death at their hands. I was well armed with an army musket and had a belt of cartridges around me, and made up my mind that there was nothing for me to do but fight. At the first shot I fired they immediately dropped down behind their ponies and began circling around and threw out . . .

[*Here Rosser switched to a different sized piece of paper, about ten by sixteen inches, and rewrote pages 6 and 7; indicating that he reread and destroyed two pages from his first draft.*]

[17] In fact, this is the route that the Northern Pacific was to follow.

. . . [T]his evident fear of my gun on their part gave me great confidence and hope and I suppose I maneuvered with great skill until we had in this way skirmished over about two miles, which occupied a very short time. Here I saw off to my right about a quarter of a mile a soldier [on foot] who had just come into view and appeared not to understand what was going on. The Indians were nearer him than I was and I could not join [——] but motioned him back [toward me]. He at once started back at a double quick towards the [infantry] escort and [surveying] party which were then about a mile back, and over a hill and out of sight from him and the Indians. As soon as the Indians saw this dismounted man they turned off from me, and went in pursuit of the soldier.

As soon as I was liberated I ran my horse at full speed to the crest of the hill, a few hundred yards ahead, where I could see the party. I was soon observed by the gallant Eckelson who seeing the Indians galloping around me, snatched a musket from the hands of a soldier and in a moment was at my side. The Indians were then very near the soldier and we rode back to help him. The few scattering fled before us, but as we raced [to] the soldier he was shot and fatally wounded, but we jumped off our horses at his side, drove the Indians away and held his body [——]. [I shot] One of the Indians [——] dead in [the] face [and in front of] of the entire cowardly savage band [which we held off] for at least half an hour, until the [infantry] company came at a run to the spot.

The soldier, for whom we risked so much to save, was Lieut. Adair of the 22nd U.S. Infantry. As soon as the company came up the savages went off to the hills whence they kept up the usual skirmishing until we finished work for the day and started to camp; then they came on the field and carried off their dead. Lieut. Adair was taken into camp in our wagon where he soon died. He is a [first] cousin of Mrs. President Grant, [and] leaves a wife and several children in destitute circumstances.

A few moments before the Indians made a dash at me, they had killed Genl. Stanley's servant who had straggled out of camp on a hunt, and when they saw me I suppose they took me for a straggler from camp also. And I retreated [away] from camp they doubtless concluded that they had me safe, and hence the leisurely manner in which they persued me. This circumstance was likely all that saved my life, for they could easily have rushed on me and killed me in the first half mile.

[The day] before they came on me, they had come upon Lieut. Crosby of the 17th who was out hunting and had killed and scalped him, and during our skirmish with them one of them [Gall] would occasionally ride out from the others and flourish over his head a long pole with a scalp on the end of it, which we knew had been taken from some of our friends, we soon learned that it was poor Crosby's.

These Indian were evidently hanging around camp for the purpose of picking off stragglers and I assure you they have proven to be a most efficient set of file closer. No man gets behind and no one asks permission to go hunting now. I hope we have seen the last of them.

We have had some very cold weather. . . .

Heart River has been frozen over for several days, the grass however is still good and our stock [beef herd] are looking well. The govt. stock have been over ten days without grain and have done well on grass. The train from Rice came in last night with forage and I have succeeded in getting 2000 pounds from the Q.M. which will take [my wagons] through comfortably well. A fresh escort will join me in a day or two and Stanley will go in, but I think I will complete the location into the Missouri River by the 25th inst.

The party has worked superbly since you left and not a day has been lost but they have worked seven days out of seven and from daylight till dark, and checked over the notes and plotted the line at night. Their zeal, devotion and energy can't be too highly praised.

Yours respectfully [etc.]

Source: Manuscript Collection, 134, L.P. (F), Minnesota Historic Society, St. Paul.

The Death of Adair (excerpt)

For days and weeks after [the O'Fallon Creek fighting] we marched and worked but saw nothing more of the Indians, and the men naturally dropped back into careless habits. They would straggle and wander miles away from camp or column hunting. Yet all went well until we had gotten back to within sixty or seventy miles of Fort Rice.

It was a bright October day, rather windy but clear and crisp. The main column, overtaking the engineers about noon, all lunched together. Then Gen. Stanley moved on as usual intending to encamp at the nearest water, estimating the distance as reported by the scouts, at six or eight miles.

After lunch the command moved on and the engineers resumed work, expecting to run the line up to the camp by night. Capt. Miner's company composed the guard for the engineers on this occasion, and Lieut. Dent Adair, though I believe not attached to the company, was doing duty with it. Taking a few "crack shots" from the company,[18] Lieut. Adair started off about 3 P.M., ahead of the survey, on a hunt. I accompanied them for several miles, and we knocked over a half dozen or more antelopes, but finding the country breaking up into rough ridges, I left the hunters and put myself on the line of survey, that I might assist Mr. Eckelson, who was conducting the survey, in preparing his map of the day's work.

As I rode leisurely along, examining a series of broken ridges over which we should be required to take the line, I observed the heads of men just over a little "swell" to my left and front, which I took at once to be the pickets surrounding Gen. Stanley's camp which I supposed was in the valley beyond. I rode on giving myself no concern about them. The line which I was examining passed along parallel to this ridge over which the heads of the men were seen [going in the opposite direction], and I soon passed on beyond them. There was a little knoll standing alone on the prairie about a half mile off, and wishing to take a last look at the country before going over to camp, I therefore loped my horse over to it, thus riding away from the supposed camp. The heads of men which I had fancied were Stanley's pickets were on the shoulders of savage Indians. They had been following along on Stanley's trail and had killed several stragglers and hunters and when I saw them they were engaged in the delectable pastime of hacking and mutilating their bodies![19]

In fancied security I sat on my horse and looked out over the broad prairie. The sun was sinking low in the west and lengthening shadows warned me that night was near. Yet the scene fascinated me, not a tree in sight and the waving grass and howling wind gave the scene one of oceanic desolation. Not a farm house or a home within hundreds of miles. A military post here and there at wide intervals where a few are sheltered, isolated and in a state of siege. Antelope curiously stared at me, and the coyote and catamount [mountain lion] lazily sulked away uttering a dismal wail as they passed in the distance to take a last look at me.

---

[18]Rosser's memory played tricks on him, as Adair was alone.

[19]This was not the case. Rosser's fabrication—seeing the Indians mutilate the bodies—can only be explained by his desire to sell the story. Other than Stephen Harris, whose body was not found until later, there were no other casualties that afternoon.

I contemplated the changes which the completion of our great railroad would bring; the market towns that would spring up with their churches and school houses. The pretty farms which would reclaim this waste of soil, and the flocks and herds which would cover these valleys and hills.

I was aroused from this revery by the sudden start of my horse. Four Indians came galloping towards me from the direction in which I had noticed the heads of men. I first thought that they were our scouts, sent out to guide me to camp. But not feeling safe in such a place holding a reception with Indians of questionable repute, I hailed them when they got within earshot, but on they came and gave me no answer. I hailed again and at the same time brought down my rifle which was a splendid regulation needle gun,[20] and fired as a caution to them. This act revealed their character and purpose. They scattered, returned my fire, and set up the usual shout "hi-I, hi-I" and "On right and left, above, below; Sprang up at once the lurking foe."

They had been watching from behind the hill for a long time—an Indian never shows more than his head from a lookout while a white man gets on the highest point he can find, and boldly stands erect showing his entire form. These savages were between me and the command under Stanley, and not knowing that the engineers and their escort were in the rear they felt sure of my scalp. They came up from behind the hills, in every direction, and in a few moments were on all sides of me. Visions of torture and death at the stake inflamed my imagination and "To my heart, the life blood thrilled with sudden start."

I was well mounted, [and] my rifle was first class. I had forty cartridges in my belt and my saddle pockets were well supplied with more. I slowly galloped back in the direction of the engineers with the savages gradually closing around me on all sides. They used their firearms sparingly, and from the backs of their ponies their aim was very inaccurate. But few of their arrows reached me and fortunately none struck either myself or horse.

I faced about frequently and fired at those nearest to me, but so far as I could see without effect, yet I kept them at a respectful distance. In vain I strained my eyes over the prairie in search of the engineers and their escort. Not a soul in sight, but howling savages intent on taking my scalp! Other feelings under such circumstances can never be described by tongue

[20]Exactly what Rosser had is unclear other than that it was a single shot weapon that used cartridges. As some needle guns were not rifled, the term *musket,* which Rosser also used, could be correct.

or pen, the vision sometimes returns to me in my dreams, but I awake so suddenly that but a short part of the tragedy is repeated. I resolved not to be captured but to make the best use of my rifle upon the savages and in extremity use my pistol on myself. No man who understands the Indian character will ever suffer himself to fall into their hands alive.

As I was passing over a low ridge I saw a soldier about a half mile off surrounded by Indians. He was afoot and using his rifle freely and standing the Indians off manfully. I realized that our chances of escape would be strengthened if we could get together, and I made a bold dash to reach him.[21] I fired right and left and succeeded in breaking through the line. When I reached him I found him to be my friend, Lt. Adair. I sprang from my horse but as I did so a tall, well painted young "buck" who besides his paint wore nothing but a "breach clout" and moccasins, with a sash of red flannel tied about his waist, galloped up within a hundred feet of us, jumped off his pony and fired. Adair and I each fired at him, and the three shots were simultaneous. Adair and the Indian fell. The Indian was shot through the heart and was dead when he touched the ground. Adair was shot through the body, and at his request I turned him over on his side, but the Indians kept me too busy to do more for him than to loosen his clothing. Soon after I reached [Adair,] Mr. Eckelson joined me and together we stood over Adair and kept up a brisk fire on the Indians until Capt. M[iner] reached us with his company and the engineers.[22]

There was no time between their first appearance and the arrival of M[iner] that the Indians could not, with a small loss of life, have closed upon and dispatched me, but every Indian fights on his own hook. A chief can order his tribe to go on the war path, and in battle he can order an advance or retreat, but if he wants anyone to rush upon a foe, and thus greatly expose himself, he can perform the service himself. His warriors may be inspired by his example, but not impelled by his command. They fight as long as they feel like it, and can then leave the battlefield without authority or disgrace.

The Indian who shot Adair was Little Red Horn who, although young, had acquired distinction for bravery and was well known along the frontier and in this attacking band he was the only one who risked his

[21]The reader should remember that Rosser's horse was a *draft* animal.

[22]There is an old saying to the effect that one should never ruin a good story with fact, an approach Rosser clearly followed in these paragraphs.

life to take a "coup." It was reported at the [Indian] Agencies that we killed eleven of the three hundred in the attacking band, but Little Red Horn was the only one left on the field.[23]

Darkness came on and Adair was still alive. We did not know how far it was to camp, and we had no means of moving him. Eckelson and I were mounted and we volunteered to ride across the prairie to camp and get an ambulance and surgeon. We rode for some distance under fire, but the Indians did not chase us. We passed on the roadside the mutilated bodies of the soldiers whom we could not recognize, and carried the dreadful tidings of the death of the gallant and popular Adair to his friends in camp. All of the mounted officers and the scouts returned with the ambulance and escorted Adair back to camp, but he died before morning.

Stanley's Official Report, August 25–October 14

The command moved slowly back on the trail, keeping pace with the Engineers, who located the railroad at a rate, varying according to the nature of the ground, of from four to ten miles daily. On the 31st of August the work had progressed to Cabin Creek the stream draining the "Bad Lands" mentioned as having been met with on the 12th of the same month. After several days' unsuccessful effort, Mr. Rosser the Engineer in charge[24] informed me that the line by O'Fallon's Creek was a complete failure, and unless we could find a better line to the Yellowstone, he must abandon the work and return home.

This matter of returning was serious only on account of sickness in the command. Cabin Creek has a dry sandy bed, 100 feet wide, but neither running water nor holes. By digging plenty of water is found, but of the most nauseous and purging kind, and at least three-fourths of the command were sick from its use. I however decided that the Expedition must not fall for want of trial, on account of the military. Fortunately on the night of the 1st of September we had very heavy rain, converting Cabin Creek into a torrent and [filling] Cabin Creek to its mouth, Thence passed over sixteen miles of beautiful table land, to the mouth of O'Fallon's Creek, returning by the route of Cabin Creek, on the 12th of September our camp was pitched near the camp first made on Cabin Creek. After tracing this stream up, south until we ran into

[23]Both numbers are unsupported.

[24]In his official report, Stanley drops the title "general," referring to Rosser as a civilian.

the "Bad Lands" of the Little Missouri, the survey was abandoned . . . and we marched to the point of our first crossing Beaver River [Inman's Fork], in which vicinity we remained six days to allow the Engineers to survey a route and tunnel through the summit dividing Beaver and Cabin Creeks. During these six days we only shifted camps for grass.

The work being completed in this neighborhood we resumed our homeward march and crossed the Little Missouri on the 22nd day of September, where Captain Read had awaited us for two days, his scouts having been out twice failed to find us. As soon as the train could be unloaded, another train made up mostly of the citizen [wagons] employed at Fort Rice was sent to that place to bring up additional forage. This train was escorted by Captain Lazelle with two companies of the 8th Infantry. The train returned by the Cannon Ball on Sully's route which Captain Lazelle speaks highly of, but found longer than the Heart River route. Mr. Rosser had expected to spend nearly one month in locating the railroad in the "Bad Lands" of the Little Missouri, but the ground proved so favorable, that on the morning of 30th of September we found ourselves entirely out of the "Bad Lands," east of the Little Missouri; the Engineers had located a route quite satisfactory, in about eight days, across this apparently difficult country.

Mr. Rosser here informed me that he would not want so large an escort, and desired that I leave him three companies, and march the rest into Fort Rice. I did not think best to accede to the proposition, but as one company of the 17th Infantry, at Fort Rice, was under orders for Fort Totten, I ordered the battalion of the 17th Infantry, under command of Major R. E. A. Crofton, to march at once to Fort Rice. The battalion left our camp on the main branch of Cannon Ball River six miles from Rainy Butte on the morning of the 2nd of October. On the evening of the 3rd camped at the crossing of Heart River 5 miles west of Heart Butte. 1st Lieut. Eban Crosby 17th Infantry, left high camp to hunt, and when about 4½ miles from camp was surrounded and murdered by 100 mounted Sioux.

The day before this murder, this same party had discovered the five Santee Scouts who had served me during the summer and whom I had sent to Fort Rice with dispatches. The wild Sioux attacked these brave fellows at sunrise at Heart Butte, and kept up the fight for nearly fifty miles and during the entire day. The Santees were well armed, had 100 rounds each, and they kept their assailants off, and came [away] with the loss of two of their horses and their blankets, clothing, and some accouterments which they dropped to lighten their horses. I recommend these brave Santees to the notice of the Commander of the Department.

Our course from the Red Butte, the point at which the route leaves the Little Missouri "Bad Lands," is nearly due east and keeps the divide between the Cannon Ball and Heart Rivers. The ground being very favorable, progress eastward was repaid, and on the evening of the 4th of October, our camp was 5 miles from Heart River. We made a march of 20 miles that day to reach wood and water. The Engineers were seven miles behind at the time we camped, Captain Miner's Company (G) 22nd Infantry being on duty as escort to the Engineers that day.

About 3 o'clock P.M. whilst I was engaged near my left picket post, the right post one mile distant signaled Indians. When I reached camp they reported from 40 to 100 Indians as having ridden within gunshot of the picket, and as soon as they had discovered the picket, had gone off as fast as their ponies could carry them. Lieutenant Badger, A.A.G.M.,[25] and one man soon started to ascertain whether these Indians had done any harm, and meeting Mr. Rosser, Division Engineer, found that the party of Indians had met my servant, Stephen Harris, and killed him, had met Mr. Rosser, and pursued him back to Captain Miner's Company, and had met 1st Lieut. Lewis D. Adair, 22nd Infantry, and wounded him, as it proved, mortally.

Lieut. Adair belonged to Captain Miner's company, and the day's march being long, and tedious, had taken one of the men's guns, saying he would go to the top of the ridge and look for camp.[26] The Indians pursing Mr. Rosser had discovered Lieut. Adair and surrounded him. As soon as Mr. Rosser learned of Adair's peril, he and Mr. Eckelson, Assistant Engineer, rode full speed to his rescue; they found three Indians near him dismounted, two of them trying to put another on his horse. Mr. Rosser dismounted and fired at the group shooting the wounded Indian through the head.

These two gentlemen stood by Lieut. Adair, and kept the Indians off until the Company came up. The Indians then got off and at long range exchanged a few shots with the company, but at such a distance that

[25]Nicholas D. Badger (1837–82) enlisted with Ohio regiment. Early 1878 retirement likely speaks to medical problems.

[26]By officially stating that Adair was looking for a campsite—one had already been selected—Stanley was protecting Adair's meager pension. Had Stanley accurately said Adair was hunting, which was personal business, any pension would have been lost.

Captain Miner did not think it worthwhile to fire but a few shots at them in return. It proved upon examination that Adair had shot the dead Indian through the chest fatally. Mr. Adair was shot with a Henry rifle, the ball rupturing the spleen. He died of this at 5 o'clock A.M. the 5th of October. Lieut. Adair was a good officer, an exemplary and very brave man. He had served with distinction throughout the War of the Rebellion, had fought in many battles, and as senior captain had commanded his regiment.

Stephen Harris (mulatto) had served in the Army, and was a very excellent and upright man. All the deaths I have mentioned at the hands of Indians come from the fatal mistake of going alone out of sight of the command, a fault my orders, my warnings, and displeasure had not succeeded in entirely suppressing. We buried Harris at the Crossing of Heart River, and brought the remains of the officers to the Missouri.

The progress of the Survey on Heart River was slow; on the 9th our forage train under charge of Captain Lazelle returned from Fort Rice, on the 10th we re-crossed Heart River at the old crossing on Major Whistler's trail, and on the 12th camped on Heart River opposite the mouth of [Big] Muddy [Creek]. At this point we were joined by Captain Pearson's Company "B," 17th Infantry, commanded by 1st Lieut. L. M. O'Brien, which had been ordered out to guard the Engineers during their finishing up of their work on lower Heart River.

In view of the disturbances by Indians I directed this company to be re-enforced by 15 men from Captain Charles Clarke's Company from Otter Crossing. I deemed this force amply sufficient for the purpose, and the keeping of this company out for 15 days would relieve the large force under my command, allowing them the advantage of seasonable transportation on the river, and at once reducing our large expenses, as I should have been obliged at least a part of my hired transportation.

On the morning of the 14th we marched for Fort Rice, reaching that post afternoon of the 15th. Having heard at 1 A.M. this morning (15th) of the Sioux [attack] upon the Ree Scouts at Fort McKean, I sent Captain Miner with 90 men of the 22nd to join Lieut. O'Brien lest he might be attacked by the Sioux. Captain Miner found everything quiet and rejoined his command at Fort Rice on the 18th. From this point the troops and transportation were sent to their respective stations as soon as practicable, and I had the honor to report myself by telegraph from this post the 23rd instant.

Our tour of service was tedious, but interesting, as our route was mostly in a country new and unexplored. The conduct of the officers and troops as well as their quality was very satisfactory, with a few exceptions to be hereafter mentioned. The courtesy and considerations for the troops, upon the part of Mr. Rosser and his assistants, was always so marked, that myself, and I believe my associates, took pleasure in giving them every facility in our power in the prosecution of their work, and left them with much feeling of good will for them.

The Yellowstone Expedition, 1872

For annoyances, trials and troubles—
For sickness, rheumatics, and pains,
Commend me to Indian campaigning
On Dakota's northern plains;
With *Stanley* the bold as commander,
His systems, precautions and rules,
His indifference to us and our comfort
And his fondness for teamsters and mules.

His staff seemed on purpose selected
From those lacking judgment and sense,
Poor *Wright's* a mere spectacled cipher,
Whilst *Ketchum's* a booby intense;
And *Lavater* would never have chosen
From so many boys wearing the blue
To *administer* duties important
One, like *Badger,* as "honest and true."

The pet "Twenty-second" was a rabble
Ne'er for soldierly qualities blamed;—
Insubordinate, ill-bred and turbulent,
For much bluster and swaggering famed;
Yet to faults of this "lightning" regiment
Stanley ever was blind as a mole,
Till the *mustangs* of *Goodloe's* battery
From old *Badger* the yeast powders stole.

A beggarly array of d——d outcasts
Were the scouts under feared little *Brush;*
Indians, half-breeds and voyageur Frenchmen;
All determined the hostiles to crush,

Like the French fabricators of chignons,
  They were all on the *qui vive* for "har,"
But when opportunities offered
  They, like Wallen in wartime, "weren't thar."

We had *"Wallen Le Grand"* to watch o'er us,
  Who more safe to in battle command?
Has he not to escape war's commotion
  Most successfully hitherto planned?
In his "*Record*" he tries to exhibit
  How he never from battle has shrank,
But when the Indians "went for" our picket
  He "went for" the opposite flank.

But for the worst of *Kindling's* troubles
  Was caused by this prairie air;
It has touched with snow his once glossy beard,
  And bleached his once raven hair;
Yet some d——d satirical cusses
  Ascribe this sad change to his years,
Or say his hair, like Mazeppa's, grew white
  From unwonted terrors and fears.

The effects of a staff elevation
  Are well marked in our Adjutant *Hyde*
The truth of the adage confirming
  How be beggars on horseback will ride;
His new gestures, his airs, and his manners—
  No doubt he considers them bland—
Are but copied from those of his patron,
  The unctuous *"Wallen Le Grand."*

We had *Tilton* of visage saturnine,
  To preside o'er our ailments and His;
Dosing all of his martial patients
  From two little cases of pills.
All the ills that humanity's heir to,
  From colic to "cute gonorrhea;"
Meet with similar treatment precisely—
  Pills astringent and sometimes beef tea.

For his hobbies the sick are neglected
  Greater interest by far does he take,
In despoiling an Indian graveyard,
  Or in cooking a fat rattlesnake.
His saddle-bags daily are filled up—
  Not with medicines, lotions or balms,
But with a choice collection of agates,
  Rocks, petrified oysters and clams.

When we started from where the Missouri
  Rolls its turbid stream on to the sea,
And the Fort Rice mosquitoes abundant
  Held o'er us their fierce jubilee;
We were bidden good bye by our friends there,
  With many a foreboding dread,
That they never again would behold us
  Till the Yellowstone gave up it dead.

But with *Stanley the Bold* as our leader
  What cared we for harassing Sioux?
When our annoyance was fitting
  Our feet to our Government shoes;
And our skirmishes, struggles or scalping,
  Who'er for a moment once thought,
When blisters on feet and on fundament
  The unwonted tramping had brought?

Take one drink of that *Cottonwood* water
  And forget all your Indian fears—
For what signifies some little danger
  To a man with the diarrhhoes; [*sic*]
Two draughts of that Sulfur Spring fluid
  Would make one regardless of God,
And give his *Zephyrs* an aroma peculiar,
  Ne'er wafted from Araby's sod.

Pass our tour lying out on the picket,
  In the chills of the night and its damps,
Then look out for pneumonia or fever,
  The rheumatics, bronchitis or cramps;

Then with pain in your bones, you're kept marching,
  And your stomach you're forced to cram
With long rations of alkali water,
  Add hard-tack and loud-smelling ham.

How we longed for the flesh-pots of Egypt,
  The languishing glance of the fair,
In whose arms we have lovingly lingered,
  As we toyed with their long silken hair;
How the champagne did brilliantly sparkle
  In the sunlight of beauty's bright smile,
As some siren with wanton enchantments,
  Our thoughts from life's cares would beguile.

How delightful one's trod in the morning,
  How pleasant at *Parker's* to lunch,
How soothing to thirsty sensations
  Were cock-tails or cobblers or punch;
And then, when the shadows had lengthened,
  And the lamps had succeeded the day,
What good fun the *Alhambras* to visit,
  Or watch *Cyprians* walking Broadway.

The pleasures of Indian campaigning,
  May go down very well with marines,
But veterans tried in the furnace
  Know only too well what it means.
This winter, when cozily quartered,
  We may think without sorrow or moan,
Of the rough trip we had under *Stanley,*
  To the banks of the far *Yellowstone.*

Source: Beinecke Rare Book & Manuscript Library, Yale University. I have modified some spelling for ease of reading.[27]

[27]The author remains unknown. While Dave Eckroth attributes the poem to then Lieutenant (he became a captain in May 1873) Augustus W. Corliss (1837–1908), "Dolly V. Arden" says that "Our Poet" was carried away by his horse and "hanging [on] to the pummel of his saddle" at one point. Given that Corliss was a cavalry officer during the war, could it still be him?

Newspaper Accounts
Sioux City Journal, October 26

YELLOWSTONE

Fort Rice, D.T., October 15, 1872

Editor Journal:

The expedition arrived here to-day after an absence of eighty-two days. We enjoyed first rate weather. We parted from the engineers at the "Big Bend" of the Heart River, two days march from here, where they have made a connection with the other line [from earlier this year]. The Indians have given us considerable annoyance since we left the Yellowstone, but the loss of life has been small on our side.

While we have killed twenty and wounded about sixteen of them,[28] they only managed to kill one of our officers and General Stanley's cook, which would not have been the case had these two been on their guard. The officer, Lieut. Adair, of the Twenty-second Infantry, was with his company (G) on engineer escort on the 4th inst., and while the column was in the act of camping, the Indians, about sixty in number, dashed over the bluffs, but out of sight and range of the engineers and their escort. The commander of the latter was almost a quarter mile in advance of his men, which unfortunately gave the Indians a chance to cut him off and prevent his receiving any aid from his men. While the greater portion of them remained on the bluffs, about six of them dashed up to the Lieut. and the foremost of them fired at him, and sent a bullet through his heart. The other six swarmed around him, and were just about to scalp him, when General Rosser rode up and, seeing the condition of affairs, discharged his piece at the savages and sent a bullet crashing through the brain of the savage who shot Adair. As soon as he fell they scattered and fled across the bluffs, and Rosser, who was well mounted, and indeed is a plucky man, followed them at his horse's fastest pace, banging away at them with his breech-loader, and causing them to duck their heads every little while when a bullet would whiz close to one of them. He says he wounded one or two of them, and I guess he did, for he looked as though he did more than wound one or two of them.[29]

[28]Both numbers are inaccurate. Throughout his letter Swipes is invariably factually wrong, a mystery for someone who was there.

[29]What Swipes meant by his comment that Rosser "looked" is unclear but nevertheless fascinating, for he is known to have become "enraged" during battle.

I believe Gen. Rosser would not harm an Indian under any consideration, except when he "gets after them," and then he only "hoops 'em up." He says he has a different way of issuing powder and ball to them than the Government, and he would respectfully recommend his way as being the best and most effective.

By this time the company came up, and finding the body of their officer but with little life in it, their rage knew no bounds, for Lieut. Adair was well liked by his men. He was a perfect gentleman in every sense of the word, and had the respect and esteem of all who knew him. Not a few of them shed tears and swore vengeance on his murderers. The soon espied the body of the dead Indian and they seized it and flung it in the air several times in their fury. They stripped it of its adornments, including blanket, revolver, carbine, etc., and left it to the wolves.

A scout was instantly dispatched to the camp, four miles off, to report to General Stanley and another company was sent back in haste to reinforce the escort. An ambulance, also, was sent with them to bring in the officer's body and, as they were returning to camp, they found the mutilated body of [Stanley's] poor cook in a little ravine, not over two and a half miles from the camp. The poor fellow, it seems, fought to the last, as the cartridge shells around where his body was lying would testify. His arms, legs, and breast were cut and hacked in a horrible manner.

All this took place about fifteen miles the other side of "Heart Butte," which is about eight days march from here. The death of this officer cast a gloom over the whole command, and Gen. Stanley felt the loss of his cook, who was very much attached to him, not a little. He had been out hunting about two miles ahead of the train. This is the same party with whom we had the skirmish at O'Fallon's Creek on the 22nd of August. I am informed [that they] were down to the Agency at Grand River[30] to get blankets, ammunition, etc. as well as liniment and plasters for their wounds, which they say they received in a fight with some other tribe, but which in reality they got from us at the above mentioned creek. They were returning when they met the engineers, which, as I should have stated, were about four miles in the rear of the main column.

When we resumed our march next day, five scouts were sent out in

[30] The Grand River Indian Agency was where the river flowed into the Missouri, just below today's North-South Dakota line. The agency was moved in 1875 to a site now covered by man-made Lake Oahu.

advance, and were pursued by the same party and one of their horses was shot dead, and it was with difficulty they escaped with their lives. The redskins are on the warpath all along from Fort Rice to Powder River, and are determined to do as much mischief as they possibly can before cold weather sets in. When we were in two days march of Fort Rice they set the prairies on fire all around, hoping that we might get burned out. And they very nearly succeeded, as we worked hard all night on the 12th inst., and much of the next day "fighting the fire," and it was with the utmost difficulty we saved the camp from being burned. The fire spread to this place and the troops were unable to save about four hundred tons of hay which was consumed, and at some of the agencies lower down [the Missouri] it has done some damage the result of which I have not learned.

While writing this I learn that the scouts belonging to this post have had a fight at the Crossing[31] and have brought in some scalps. Lieut. Crosby of the Seventeenth Infantry was killed and scalped at the latter place, and two soldiers wounded in a brush with them. The Indians were routed. The Indians have suffered considerably from the day they attacked us at O'Fallon Creek up till the present. They have been severely walloped very time they have shown fight.

The route that has been surveyed for the railroad is a very good one, and for the most part runs through valleys well watered and wooded, and affords good facility for settlers. The land is good in most places, especially along Heart River and O'Fallon Creek. I will here state, if not out of place, that a better commander than General Stanley, for an expedition like this, it is difficult to find. He is cool and collected when any danger threatens and is particularly attentive to the comforts of his men and animals. He is a good soldier, and never puts his men to any unnecessary trouble. The engineers, too, are jolly fellows, and seem to enjoy a little fracas with the redskins, occasionally. Well, the expedition has returned, and we are not sorry for it. The command breaks up tomorrow, or next day, and all go to their respective winter quarters.

Yours very truly, N.S. [Nibsey Swipes]

---

[31]Swipes meant Fort McKean.

New York Herald, October 28

WAR WITH THE REDS
TWO BATTLES WITH THE SIOUX NEAR FORT MCKEEN
FOURTEEN INDIANS KILLED AND WOUNDED—
FIVE MILITARY SCOUTS AND TWO OFFICERS OF
THE STANLEY SURVEYING EXPEDITION
MURDERED BY THE SAVAGES—
LIEUTENANT ADAIR, A NEAR RELATIVE OF
PRESIDENT GRANT KILLED.

Edwinton, D.T., October 12, 1872

Owing to the disposition on the part of the manager of the telegraph line east from the Missouri River at Edwinton to suppress all news of hostile Indians, it has been impossible to communicate exciting intelligence of the doings of the Sioux recently, on the west bank of the river, in the "land of the Dakotas" [emphasis added]. A few days since the hostile warriors, to the number of 150,[32] screened themselves in a ravine about a mile west of Fort McKean, watching for the appearance of any game in their line. They had not long to wait. A couple of Arickera scouts, employed in the military service, approached apparently without observing the enemy. They were immediately surrounded, but stood their ground and fought bravely, and as soon as the firing was heard at the post fifteen or twenty more scouts joined their comrades. The garrison was aroused to arms, but before any assistance could be rendered by the troops to their allies, the Sioux had departed out of reach. It was a fierce and heroic fight on the part of the scouts, two of them being killed and one wounded, while eight of the enemy were killed and wounded.

Before this affair had fairly blown over news came of the death of Lieutenant Crosby, 17th Infantry. He was in camp with the engineers of the Northern Pacific Railroad, near Heart River, about fifty miles from Fort Rice, belonging to the escort of two companies of infantry, who were on their way in from the Stanley surveying expedition. The Lieutenant had wounded an antelope which came near the camp, and

[32]Virtually all whites disagreed about the number of Sioux, with ranges varying from fifty to two hundred. If there was only one band of warriors, led by Gall, fifty would be the maximum. However, if Sitting Bull joined him—as some believe—then the minimum would be a hundred.

in opposition to the advice of friends he pursued it over a bluff but a few hundred feet from the command when he was surprised by Indians, who were waiting for just such a chance as this.

KILLED, SCALPED, AND HORRIBLY MUTILATED

It was done so stealthily that it was not known by the officers of the camp, till being absent longer than they thought necessary, a party went in search to find him as described. A few shots were exchanged with the barbarians without sanguinary results, when they hastily retreated from the soldiers. Lieutenant Crosby went through the war losing an arm in the service, and he was an esteemed officer in his regiment.

A day or two later still another murder was perpetrated by the savages. Lieutenant Louis Dent Adair, Twenty-Second Infantry, belonging to the Yellowstone expedition, was riding besides General Rosser, Division Engineer Northern Pacific Railroad, when the two were attacked by Indians and the Lieutenant was shot through the body. He returned the fire after being wounded fatally, and it is reported, killed the Sioux who shot him. He lived nearly two days after receiving a mortal injury. Adair leaves a wife, two children, and numerous friends to mourn his death. He was a near relative of the President.

The surveying party expected to arrive Fort Rice today, the 15th inst., and it is believed not to have been an entirely successful expedition so far as permanently locating the exact line of grade for the railroad. No work will probably be done next season west of the Missouri River more than that of the engineers'/at least, the question is in abeyance. Many of the leading officers of the road have gone to Puget Sound, where everything is progressing satisfactorily and rapidly. A line of steamers, to run in connection with the railroad, is to be put on in the Spring, plying from the end of the track up and down river; and the carrying trade will be large, making a shorter and quicker route east from Montana, via Fort Benton, thence by steamboat to Edwinton, on the North Pacific Road, to Lake Superior, probably diverting a great deal of the freight and passenger traffic from the Union Pacific Road. This line of steamers will be owned by and under the direction of the North Pacific Company, from which there must come very handsome revenue, as it will be economy for the government to ship all military and Indian supplies by this route for all the Missouri River posts and reservations in the Northwest.

Yankton Herald, December 12

This tongue-in-cheek letter from a soldier writing under the then-humorous pseudonym of "Dolly V. Arden"[33] *is particularly interesting in its attitude toward David Stanley. More than a month after Stanley's second drinking binge, the writer still clearly retained the highest regard for him.*

LETTER FROM FORT SULLY
BY DOLLY V. ARDEN

Fort Sully, D.T., November 25, 1872

Editor Herald:

It is not unlikely that you, and probably some of the readers of the *Herald*, have wondered what has become of Dolly and why she don't further inflict the public with her scribblings.

The fact of the matter is she was absent as the "Daughter of the Regiment" with General Stanley's Yellowstone Expedition where, in the multiplicity of other cares, she was unable to devote any time to the enlightenment of the human race.

Her career with the expedition was quite romantic, and only her extreme modesty prevents the world from knowing how very often she came near falling a victim to "Poor Lo's" passion for Chignons.[34]

Other and less interesting pens must record her deeds of daring.

The sights, scenes, and incidents of the trip would make an interesting letter, but I feel my inability to do the subject justice; consequently I will not attempt it.

The meeting of Genl. Stanley and "Gaul" on the banks of the Powder River would be a glorious subject for a painter.

There stood our brave commander, alone, unarmed, fearless. Directly opposite in a clump of timber not twenty yards distant, stood "Gaul" the implacable chief of a hostile tribe. [B]ehind every tree and shrub, as was afterwards ascertained, was hid a bloodthirsty Sioux covering our beloved General with his deadly rifle, waiting for the signal from their

[33]Dolly Varden was a popular character in Charles Dickens's 1839 novel *Barnaby Rudge*. The Dolly Varden name was given to an 1869 Nevada gold mine, a popular early 1870s style of woman's clothing, and even a species of northwestern American-Canadian trout.

[34]A chignon is a roll or ball of hair at the back/nape of the neck, a style in use for almost a hundred years by 1872.

treacherous chief. Nothing but the "Old Man's" undaunted bravery, for which they have a profound respect—saved his life.

When the Genl. told him in thunderous tones, without a tremor in his voice, "The railroad will go through here *in spite of you*," "Gaul" fairly cowered, and retired, after humbly begging something to eat . . .

[*Section on Adair's death omitted.*]

The trip was replete with laughable incidents from the time we parted company with Mrs. Maloney and her beer at Rice till our return.

Everybody, from "Our Poet," hanging to the pummel of his saddle as his horse galloped over the plain, down to "Johnnie-come-lately," who wanted to discharge a twelve pounder [Napoleon gun] while it was being held up by the soldiers, contributed to the general amusement.

I will not occupy your space by entering into details, suffice to say everybody enjoyed themselves hugely, Nibsey Swipes of the *Sioux City Daily Journal* to the contrary notwithstanding—and no one more.

Surveyor Edward C. Jordan's 1906 Account

Edward C. Jordan (1846–1935) was a college-trained civil and railroad engineer, one of only two (the other was Montgomery Meigs) associated with the eastern surveys. This document represents the transcription of a presentation probably given at the 1906 surveyors' reunion in Minneapolis. While some of the comments seem rather farfetched, and there are interesting mistakes in Jordan's memory, the fact that Jordan was speaking to other surveyors gives his account a higher likelihood of accuracy. Sections dealing with non-1872 survey events have been removed.

[*At least one page missing before account continues.*]

[At the beginning of October] innumerable heads of antelope were running right and left of us. The desire to shoot one often made us stray off a quarter of a mile or so behind a hill. The rules were so strict that no private was allowed this privilege, but an officer or an officer of the Engineer corp. was often rash in this way and the result was misfortune overtook us.

We started out one morning with the usual escort of infantry, 150 men on the line. The larger part of the expedition with its wagon train marched on ahead 16 miles and went into camp. After luncheon, Gen.

Rosser the Chief Engineer left me while he went on to see what kind of a railroad country there was in advance.

Lieut. Adair, who was a nephew of Mrs. Gen. Grant, started out on foot a little on one side to get a shot at some antelopes. Rosser had ridden about three miles when about 40 Indians appeared in front of him. His first thought was that they were our own mounted scouts, but after a few moments he saw that they were hostiles. They fired and he returned it showing them that he was armed. He commenced to retreat, but did not dare to let them see how slow his horse could run from the fact that it was one of the team horses, his own fleet horse being saddle sore. The Indians did not hurry as they thought they had him sure. They had seen what they supposed to be the whole train go by and they were between him and the train. They did not know that a surprise was in store for them.

In the next depression of ground the Indians when they came in sight again had dropped out four or five who really galloped off on one side out of sight intending to get ahead of the Gen. and come up in front of him and then whoop him up. The Gen. was retreating as fast as he dared to thinking that he never would come in sight of us when looking off on the right he saw a solitary man and rode towards him and motioned him back.

Just at that moment four Indians came up from the other side of the hill as they supposed to head off Rosser when they stumbled right on Lieut. Adair and shot him and jumped off their horses to scalp him. Eckelson and myself who alone had horses galloped to a little knoll, saw Rosser, gave the alarm, and then galloped to the rescue. Rosser, who was a beautiful shot, hopped off his horse for a steadier aim and shot the foremost Indian as dead as a herring. The rest of the 40 Indians charged down towards us not yet having seen the 150 men. It was very frightful for a few moments until the Indians caught sight of our men, then they scattered quick but not before we wounded two of them. It was a poor swap—a dead officer for a dead Indian Chief. It is against the rules for a soldier to remove a scalp but if that Indian fell back into their hands unscalped he would have been buried with honors, otherwise disgrace. Not being a soldier I removed his scalp lock which I now have, and on the sly a soldier took the top off his head.

We commenced our mournful retreat towards the train and had gone but a few miles when we came upon the dead body of Caesar [Stephen Harris], the Negro servant of General Stanley. [He had been] shot and his body [was] sticking full of arrows but not scalped. They do not scalp Negroes. It was a long anxious retreat, fired at as we were every step until we reached the main command, to be told there of a similar disaster. It seemed that Lieut. Crosby had dropped behind [Crofton's wagon] train to get a shot at some antelope when these same Indians came upon him from a side ravine, shot, scalped and frightfully mutilated him and it was his scalp that they had been waving at us. It was a sad day and no more antelope were gunned after [that].

A few miles further on we came upon a place where a large Indian village had formerly existed. There were several graves and as I had promised a specimen to a young doctor I availed myself of the opportunity. The manner of burial is odd. They place four poles in the ground and lash a platform to the top. Then they dress the body in all the finery he was possessed of, lash on him a buffalo skin and place on top of this platform to dry. The one that I obtained had been drying over 20 years. He was a big brave. Nothing but the skeleton remained enclosed in what must have been an elegant buckskin suit beautifully wrought with beads and pinned upon his breast was seven scalps, three of them blonde in color showing that they must have been white scalps and perhaps all of them. There were half a dozen pewter spoons and etc. which he probably obtained at the same time he did the scalps by murdering settlers or robbing an emigrant train. One other article that I found was a red pipestone pipe. This material is very highly prized by all Indians and is only to be found in one place in the whole world.

It has a beautiful history. Where it is found in Dakota [is] near the Minnesota line due west of Saint Paul [and] was neutral ground. No matter how much at war one tribe was with another when they arrived on this neutral ground they buried the hatchet and smoked the pipe of peace. On that ground the Great Spirit had declared that they must come together as friends. Excavations are found of unknown antiquity where former generations have conformed to this custom of obtaining the pipestone. But the warlike Sioux have broken up the custom and now no other tribe obtains it except through purchase.

A few days more now, necessarily uneventful, brought us back to the Missouri where the expedition disbanded. Our friendly Indian Scouts whose timely warning and knowledge had probably saved our lives were to return to their reservation. . . .

Below us on the river was a large band of Sioux not as yet hostile in array against the whites but they came up before our fort and sent word that they meant no harm to us but they meant to kill every one of these renegade Ree Scouts that went with us on the expedition. Our scouts were brave and reckless and having plenty of government ammunition, 40 of them dashed out against more than double their number. The fight was entirely on horseback and of the most thrilling nature as they dashed back and forth sometimes right under the stockade of the fort. There was a great deal of firing but they screened themselves so well behind their horses that for [——].

Finally one of our scouts was shot and fell from his horse who galloped into the fort.

It was a long anxious retreat. [There] was a great whoop on the part of the Sioux and a party of them galloped up to scalp him when a 15-year-old son of the scout who had fallen was seen running as fast as his bowlegs could take him straight to his father. The Sioux ignored him but he reached his father first, seized his carbine and shot two of them dead, wounded a third and with the help that arrived captured the scalps and bore his father's body into the fort. The squaws that night carried on the side show of wailing but there was a grand jamboree and scalp dance among the Indians.

I had intended to give you more of a description of the odd and peculiar country, its wonderful rivers, its deep canyon, square miles of country covered with the petrified remains of forests, fossil remains of animals, the mastodon fossils, shells that you crush at almost every step for miles of travel in beds of old lakes, but such a description would be a vast labor and then far from being adequate to the subject, but an account of passing events from an eventful experience in a new country is easy except to determine where to stop.

Source: Edward C. Jordan, personal papers, held by collateral descendant Mark H. Jordan of Portland, Maine.

The Western Survey Returns

Major Barlow's Journal, August 21–October 1

AUGUST 21.—THE gale was less severe during the past night. The morning was cool, and a good day's march was made towards the divide separating the Yellowstone from the Musselshell. Our course was about north 80° west. This direction was taken in order to reach a low point in the range, where the guide said we would find small springs of water. The character of the country is the same as that passed over yesterday, being a dreary, barren, and alkali waste, broken up into deep and tortuous ravines in which occasional pools of warm alkali water are found. About twelve miles from last night's camp we reached the hills forming the divide. These hills are sand-rock, with pine trees of a scrubby sort, growing upon their upper slopes. The only water that we could find was in a few pools, where buffalo had wallowed, and it was warm, muddy, and extremely distasteful. Enough of this water was obtained for the men and a small part of the animals, but most of the latter were obliged to do without.

AUGUST 22.—AN early start was made (5.30 o'clock) in order to ascend, during the cool part of the day, the steep hills leading to the plateau some three miles above. The road selected, winding up the spur on which we had encamped, required a good deal of labor, with picks and shovels, before it was passable, and nearly four hours were consumed in getting the train up. These hills are composed of disintegrating sand-rock, much worn by the elements and standing in many detached cliffs of varied forms of beauty. On reaching the summit, a broad plateau extends off northward for several miles. Here, at my suggestion, Colonel Haydon established a base-line, and from either end triangulated to various prominent landmarks, some of which had been located from his line of survey.

From this base the Yellowstone Range, Pryor's Mountain. Bull Mountain, Judith, and Crazy Mountains are visible. Thus we have located these ranges with considerable accuracy. The hills we ascended this morning do not form a ridge, but are simply the slope of a vast plateau, about ten miles in width, lying between the two rivers, and elevated several hundred feet above the valleys. The surface of this plateau is an undulating prairie, destitute of trees of every description, and nearly so of water. By bearing away nearly due west about seven miles, the guide [Mitch Boyer] led us to a peculiar sink-hole lined with clay, and containing sweet though muddy water in abundance. The guide says this water remains permanently, even if all the springs in this region dry up. It is simply a pond, or rather a ditch, seventy-five yards in length, with a hard clay bottom, and although our animals must have drank three thousand gallons from it the depth of the water was not perceptibly affected.

There was no wood here, and before going into camp it was desirable to reach the northern edge of the plateau, where both wood and water would be found. The whole plateau had just been alive with buffalo, and fresh meat will now be plentiful. The weather, however, has been too hot to admit of hunting the buffalo with anything like comfort. About six miles from our last watering place we came to the edge of the plateau leading down into the Musselshell Valley. These hills are very picturesque, and the view as we reached the edge was splendid. To the north lies the Judith Range, dark in color, except where slightly spotted with snow; nearly west is seen the Crazy Mountains, still more snow-covered, while between is the diversified Muscle-shell Valley, with woody ravines and grassy slopes, on which thousands of buffalos were quietly grazing. This valley is much more inviting in appearance than the Yellowstone.

The heavy train moved slowly to-day on account of the scarcity of feed for the animals during the past two days, and it was dark before the rear came up. We camped among the foot-hills of the northern slope of the great middle plateau and found water tolerably abundant in springs, some of which were very fine and cold.

AUGUST 23.—A late start was made this morning in order to allow the nearly-famished animals to get the benefit of the good grazing at this camp. The route continued down the northern slope of the divide, on which we had encamped last night, and proved very winding, this region being badly cut by ravines. We found quantities of game, including buffalo, antelope, and sage-chickens

The nature of the soil changes as we approach the valley of the Musselshell, and becomes more barren though less hilly. The grass has been completely eaten off nothing being left but the universal prickly-pear to relieve the desolate condition of the ground. Very little water was discovered, and that little proved to be very poor. After traveling ten miles we reached the river, which at this point is clear and sweet with considerable current. It is about twelve to sixteen inches in depth, and forty feet in width, which seem very insignificant when the immense area tributary to the stream is considered. It is extremely crooked, with steep bluffs jutting from the adjacent highlands on either side, with enclosed bottom-lands of considerable extent lying between.

These bottoms are well wooded with cottonwood and willows.

The grass is not abundant, though much more so than upon the hills above. The weather has been extremely warm for several days, probably upward of one hundred degrees, causing great suffering, both among the men and animals. The main column did not reach camp until sunset, though the distance traveled measured only ten miles. When we reached this camp buffalo were observed in vast herds in all directions, but the injudicious hunting by several members of the command soon drove them far away, and for some days they will probably be extremely scarce.

AUGUST 24—It was decided to remain here, perhaps days, to allow the engineers to examine the country several miles below. The animals need rest and a chance to graze and recuperate after the fatiguing marches of the past several days.

AUGUST 25.—We spend to-day also at this camp. No duty is required except the necessary guard. We enjoyed a slight rain-fall last evening, which has cooled and refreshed the atmosphere wonderfully. Preparations are being made to escort the surveying party down the river about twenty miles, whence they propose to commence their survey up the valley of the Musselshell to the Missouri Forks. The escort for this purpose will be the whole cavalry force, with three days rations and a blanket for each man. The infantry will be left here with the entire wagon-train parked for defense. The cavalry and surveying party will be out two nights.

AUGUST 26.—The object of the expedition down the Musselshell River is to ascertain the location of the " big bend," as it is called, or the point where the stream turns from its eastward to its northward course. A sharp elbow is formed there, and its position, as represented by the guide, materially differs from that laid down on the maps. This bend will serve as a starting-point for the survey of the railroad line along the Musselshell back toward the Gallatin Valley. The party, with its packs, started at half-past seven o'clock and, taking a course nearly due east, passed through the underbrush and trees on the bottom-lands and ascended the ridge beyond. This course deviated considerably from the river, as the latter bears off to the southward. The country is open, being a high, rolling prairie, intersected by ravines, so common in this region, and destitute of nearly all vegetation except the prickly-pear. The valley of the Musselshell in this section seems wholly worthless except, perhaps, for grazing. Continuing onward the elevation became greater, and also our distance from the river, whose valley was soon obscured by intervening heights. After about four miles we reached the highest ridge, which must be seven hundred feet above the bottom-lands of the river. The guide told us that we would now descend into the valley of the river again, and that we would find ourselves in the Musselshell beyond the "big bend," and that its waters would be found running north from there onward to the Missouri.

The guide's information seemed correct, for, about three miles farther on, we saw that the course of the stream had decidedly changed to the northward. A great error is apparently discovered to exist in the maps of this region. Toward the east and northeast a high, rugged, and unbroken range of hills extends far off to the northward, effectually barring the progress of the river eastward, while along the foot-hills of its western slope the valley of the Musselshell River could be distinctly traced for several miles. Careful triangulations and other investigations were made by the surveying party to verify this conclusion, and it was then determined to pursue the search no farther, but to take this spot as their initial point for the survey of the Musselshell westward to the mountains. A survey of the country eastward will be attended with great difficulty. The barren, dry ridges present a most forbidding aspect, being the commencement of the "bad lands," which extend from this point to the Missouri and Yellowstone Rivers. All the information that can be obtained goes to prove that, on account of the rugged nature of the country, and the almost entire absence of water, it would be next to impossible to force a passage through with a heavy wagon-train.

The Musselshell diminishes in volume as it proceeds onward, its waters being gradually absorbed into the gravel and sand formation. Water falling upon or flowing through this soil is apt to disappear, and springs are exceedingly scarce. The bluffs in this vicinity are often bald and rocky, the rock being the usual sandstone of inferior quality, sometimes resting upon beds of clay and supporting sand, clay, or gravel. Several quite extensive strata of lignite were observed. After completing their work of triangulation, the engineers assumed a "bench mark" near the river and began running their line westward along the valley of the stream. As our distance from camp was not great, it was decided to return for the night, rather than keep the command separated, and accordingly at 6 o'clock work was discontinued, and the force started for camp. About two miles of survey had been accomplished. Darkness came on before we reached our destination, and the ride seemed long, gloomy, and fatiguing. Our camp is located upon a high knoll on the north bank of the river, the surrounding country being a flat on which there is sufficient grazing for the animals for the present.

AUGUST 27.—We still remain in this our first camp on the Musselshell. The engineers were out surveying from the point where they left off yesterday, and made about four miles additional. Their line was selected along the river, following its various curves, and remains on the north bank of the stream.

AUGUST 28.—The weather is beautiful, though very warm in the middle of the day. The engineers made good progress, and continued their survey to within a mile of camp.

AUGUST 29.—Camp was moved to-day five miles up the stream, and established just below a high, rocky point, said to be the beginning of a considerable canyon. The route was through the river-bottom, the wagons finding no obstructions of any importance. The general features of the country remain the same as around our previous camp, viz: rolling hills and valleys stretching away to the north and south of the river, commonly destitute of vegetation, while the narrow river-bottom presents a green and attractive appearance on account of the luxuriant willow and cottonwood trees growing near the water. We found the stream much swifter and cooler than at the previous camp. The engineers brought their survey up to our present camp, the distance being seven miles.

AUGUST 30.—We remained in this camp to-day, the engineers continuing the survey some five miles further up the river. The day has been most oppressively warm and dry.

AUGUST 31.—Camp was moved to-day, and a march made by the train of about ten miles. The stream was crossed four times to avoid high bluffs. By this course a pretty good road was selected. The surveyors continued their line on the northern bank of the river over very rough and broken country, encountering bold, rocky cliffs, and sharp, deep ravines. Much skill is evinced by Colonel Haydon in the selection of his line. Some three miles above the point where he left off surveying yesterday he reached the valley of a considerable stream coming in from the north, heading in Judith Mountain, and called by the guide "Crow Wing Creek." This stream, together with the rugged country along the north bank of the river, determined Colonel Haydon to cross to the other side. Just above this point the river passes through a picturesque canyon, with perpendicular sides of sand-stone rock cut into headlands of various shapes by lateral ravines. The train now passing along I joined it. The train soon led down one of the ravines just mentioned and entered the canyon, which soon grows wider, spreading out into a broad, grassy meadow. We now halted and made camp in the high grass near a large grove. The weather has turned much cooler today, rendering the march less oppressive than was anticipated.

SEPTEMBER 1.—A rainy day in camp. No movement and no survey.

SEPTEMBER 2.—The rain fell last night in torrents continually and the night was inky dark. About 3 o'clock A.M. the report of a rifle was heard, which startled the whole camp. The feeling of being roused in a pouring rain to repel an Indian attack was not agreeable. I remained in bed waiting for another shot. None came, however. The sentinel who fired stated that he saw and fired at a moving object, which he thought might be an Indian on horseback. It was possibly a wolf. The day has been cloudy and damp, though little rain has fallen since morning. The surveyors spend the day in camp rather than risk a wetting by going out on the line.

SEPTEMBER 3.—It was raining again last night. As no progress was made on the survey yesterday the camp was not moved today. The surveyors went back on the line, making about three and a half miles. The first half of the day was cloudy. At noon the weather changed, and the sun came out bright; there is no breeze, and the air is very sultry.

SEPTEMBER 4.—A march of about six miles was made to-day up the valley of the Musselshell. Soon after leaving camp the high bluffs inclosing the valley became less bold and precipitous, and at the same time gradually closed in on either side, rendering it necessary for the wagons to leave the bottom-lands and ascend the terrace immediately above. The river has about the same width and depth as when we first met it, and winds in a tortuous channel from side to side of the valley, here and there running at right angles against vertical precipices of sandstone rock fifty to seventy feet in height. Cottonwood trees, dense thickets of willows, currant and rose bushes border its banks.

The grass in the bottom is green, and in pleasant contrast with the short and withered buffalo-grass on the hill-sides. I noticed a fine variety of sandstone this morning, which will perhaps be valuable for building purposes at some future time. Along this stream, imbedded in strata of argillaceous rock, are found numerous specimens of fossils. It is from these shells that the river derives its name. From a high butte, near our route, I obtained an extensive view of the country adjoining the valley. The center of the Judith Mountain is nearly north of this point; the Crazy Mountains a little south of west. The intermediate country is much broken, and made up of barren hills of local drift, and some "bad lands," cut by deep and sharp ravines. Very little vegetation is perceptible at this season.

The surveyors are examining a line along the edge of the bluffs on the north side of the river, with a view of rising from the river-valley to the table-land above, in order to avoid the frequent crossing and recrossing of the stream, which, on account of the numerous bends, would otherwise be necessary. We camped upon the bluff on smooth ground, but with the prickly-pear in most unpleasant abundance. The river-bottom affords grazing for the animals, and wood is plentiful.

SEPTEMBER 5.—A violent rain-storm began in the middle of last night and has continued all day, keeping the surveying party close in their tents, and making it necessary for us to remain here at least another day. This season seems to be unusually rainy. Our tent-stoves are exceedingly comfortable these cold, wet days, and save us from suffering. Great excitement was caused to-day by the appearance of a citizen from Bozeman, who had been sent out with the mail for this expedition. He recounted a terrible experience. Leaving Bozeman on the 24th with a large mail-bag and several days' supplies and an extra horse, he struck down the river, expecting to find us near the Big Horn. When about opposite that point he was attacked by a small party of Indians as he was about camping, toward the morning of the 29th. The Indians came upon him so suddenly that he had hardly time to escape with his horses, leaving all his supplies, including his provisions and the mail. The Indians followed, but having two horses he was enabled, by changing from one to the other, as each became tired, to outrun them, and finally made his escape across the divide to our trail on the Musselshell. He reports that he observed from a high ridge that the country across the Yellowstone was literally covered with Indians. He followed our trail without food for four days. At last he found a few crackers [hardtack] at one of our abandoned camps, and they revived his nearly exhausted strength. He reached our present camp at three o'clock scarcely alive.

Among the few letters which he happened to have in his pocket was a request from the civil authorities for the arrest of some horse-thieves, who were supposed to be among the citizens accompanying our party. Two of the persons named were known to be present, and were at once arrested. The night set in dark and rainy, and at about half-past 8, Gorman, one of the prisoners referred to, slipped away from the guard and disappeared in the darkness. Three shots from the sentinels created the impression in camp that the Indians were attempting an attack; and until the cause of the firing was ascertained the camp was thoroughly aroused. The clouds broke away in the night, and clear weather for the morning was anticipated.

SEPTEMBER 6.—The morning was clear, and with a strong northwest wind blowing. The surveyors were out in good season, but severe showers soon came up and compelled them to return to camp before much had been accomplished. The ground is so saturated with the rain that it may be impossible to move the train for a day or two. The sharp barking of the coyotes, mingled with the deeper howl of the gray wolves, is nightly heard about our camp, more particularly when we have remained for a few days in the same place.

SEPTEMBER 7.—The surveyors made four and a half miles to-day, but as the low ground has become simply impassable for the heavy train we are still compelled to remain at this camp. In the afternoon the sun came out very warm, and the rain seems over for the present. In view of the scarcity of grass here it has been determined to move tomorrow if possible.

SEPTEMBER 8.—A cold, clear night, with some wind, has dried up the mud, and an early start was made. We crossed the river opposite camp with the train, and ascended the rolling hills on the other side. Getting up these hills, and crossing an ugly ravine, caused considerable delay. After passing these difficulties the road proved good, and about four miles from camp the train recrossed the river, and continued along the northern bank. Here the valley widens out quite extensively, while the bluffs are much more gentle and softer in their outlines than in that part of the valley through which we have just passed.

There is little fertility in the soil, except in the low ground, the upper benches being usually composed of broken rock and gravel. The country appeared very beautiful along our route to-day, reaching back from the river in both directions, and diversified with winding ravines, in which are usually found a stunted growth of pines, and, in some instances, cherry and rose bushes. The former horizontal strata of sandstone are here frequently found thrust up at various inclinations, often vertical. The long ridge of the Judith Mountains seems to lie parallel with our route, directly north of us, while the broken and snow-clad peaks of Crazy Mountains bear a little south of west.

The wagons made about ten miles, and went into camp in a fine meadow near the river, in the vicinity of a spring of water almost icy cold. Just before reaching camp a thunder-storm came up, and has continued during the remainder of the day, though accompanied with but a slight fall of rain. The surveyors continued their line some five miles, nearly reaching our present camp.

SEPTEMBER 9.—After a terrible rain-storm, lasting nearly all night, the morning broke clear and cold. As the surveyors had nearly reached camp the previous day, it was decided to move camp a few miles farther on. The valley grows wider, with sipping hills of gravel, which gradually rise to a broad level plateau, whence rise ridges stretching off toward the Judith on the north, and toward the divide between the Yellowstone and Musselshell on the south. A few miles from camp this morning the clouds in the west became black, and very soon the rain again fell in torrents, drenching everybody. The train moved on, cutting deeply into the moistened earth; and it was with great difficulty that the wagons were dragged through the low places. The river seems to increase in depth and rapidity of current, though its width remains about the same. After a march of six miles it was decided to encamp in the river-bottom, in a sage-brush flat. The rain now ceased and sun came out, gilding the heights of Crazy Mountain and the Belt Range, now quite plainly visible to the north.

SEPTEMBER 10.—Camp was not moved to day. We are nearly surrounded with bluffs, which here are gently rounded and smoothed off in soft-curved surfaces, connecting the upper and lower terraces. These slopes are destitute of tree and bush, but yield a scanty growth of buffalo-grass, while the upper terraces are usually covered with prickly-pear. The weather has been cold, fires being necessary all day for comfort.

SEPTEMBER 11.—The train made a march of twelve miles to-day, moving across the north side of the river and continuing nearly all the way over a perfectly level road. I rode up to the crest of the ridge near Judith Mountain and found a broad table-land stretching off to the base of the mountain. From this plain I could see the tops of some of the . . . Yellowstone peaks, while the Crazy, the Judith, and the Belt ranges were all in full relief. We traveled to-day directly between the Judith and Crazy Mountains, and passed a large stream heading in the latter range. One or two small creeks were crossed, heading in the ridge toward Judith, but none worthy of being named. Camp was made on the first terrace, some thirty feet above the river-bottom, with little fuel, however, within reach, as the trees along the river are becoming scarce.

SEPTEMBER 12.—We remained at this camp to-day; the country improves in beauty and fertility as we proceed up the river. From our camp Judith Gap is visible to the north of us, and the Judith and Crazy Mountains are on a line with our present position. The soil here seems very fertile, yielding an excellent and abundant growth of bunch-grass; there is much gravel, and irrigation would probably be necessary to insure successful crops.

SEPTEMBER 13.—The journey up the Musselshell Valley was resumed to-day; the road was excellent, mostly on the first terrace, above the river-bottom, where the soil is dry and hard, and the country passed over to-day was very little cut by ravines. Some seven miles from camp we crossed a broad meadow, through which winds from side to side a small thread-like stream, whose banks are destitute of trees of any kind. This stream is called Haymakers' Creek. From this valley the road led up a steep bluff, and continued for a short distance upon the bench above and then descended to the river-bottom. Another stretch of level road was followed by an ascent to the bluffs again. A few miles further on a second small stream was encountered, where the train returned to the river-bottom. A camping-ground was selected about two miles from the forks of the Musselshell, after a march of fourteen miles. Two small streams, named, respectively, Little and Big Elk, were noticed on the south side of the river, coming in from the base of Crazy Mountains. We are now nearly surrounded by hills and mountains, and our survey up the Musselshell is almost completed.

SEPTEMBER 14.—We remained at this camp today. The surveyors went back some miles to relocate their line by crossing the river to the south side.

SEPTEMBER 15.—Camp was moved up the south fork of the Musselshell, three miles above the junction of the two forks. The surveyors continued their work, and brought their line up to within a mile of the new camp. From one of the high outlying peaks of the Crazy Mountains I obtained a fine view of the surrounding country, and afterward ascended the stream as far as the mouth of a considerable tributary joining it from Crazy Mountains. This stream is nearly as large as the main fork above the junction, and both are now simply mountain brooks, clear and cool, and of rapid current; the descent being about fifty feet to the mile. A mail-carrier came in from Camp Baker, bearing dispatches for the commanding officer and bringing some newspapers.

SEPTEMBER 16.—Camp was not moved to-day.

SEPTEMBER 17.—The engineers are still surveying below camp, and we remain here until tomorrow. I rode out along the valley of the North Fork toward the mountains, to the vicinity of Fort Howe, now abandoned. By this valley wagons are often taken from the valley of the Musselshell to Camp Baker, some forty-seven miles distant from the forks.

SEPTEMBER 18.—The camp was moved about five miles up the valley of the South Fork, the railroad survey continuing along a perfectly practicable line. Four miles above camp, or seven miles from the forks, the survey crosses the tributary, coming in from Crazy Mountains, and occupies the high, rolling plateau beyond. Copper abounds in the rock in this vicinity, and fine granite protrudes through the sandstone formation, the strata of the latter being tilted up in a nearly vertical position. Several lines of stone mounds, supposed to mark an old Indian trail, were found to-day in the valley, extending over the hills and across the ravines with great accuracy.

SEPTEMBER 19.—The train moved up the South Fork. After following the open valley for about five miles the road enters the mountains; the stream becomes narrow and rapid, with numerous beaver-ponds and dams. Timber almost entirely disappears along the valley, but the hills have a fair quantity of pine. The road proved rocky after entering the mountains, large broken fragments of granite being strewn along the way in great profusion. Camp was located four miles from the entrance to the mountains, after a march of nine miles in a small level valley completely surrounded by high hills and mountain-peaks. A superb dinner of black-tail deer, &c., amply repaid us for our day's fatigue.

SEPTEMBER 20.—A cold rain came on last night, followed by a slight snow-storm, which gave a white covering to the ground before morning. This soon disappeared, and the day has been quite comfortable. The stream gets much smaller, and divides into a number of little creeks of cold water just above the camp. The country here is very wild and broken, though the engineers are sanguine of finding a practicable passage for their line.

SEPTEMBER 21.—The country on all sides presents a beautiful variety of mountain scenery, though but scantily timbered. Pure, cold streams wind through the various ravines, fringed with clusters of willows. Cottonwood has disappeared. Pine, to some extent, is found on the upper hill-sides. The sandstone rooks of the lower valley are tilted up nearly vertical, while the mountains are formed of solid granite well exposed.

SEPTEMBER 22.—A drizzling, cold rain set in today, which, toward night, changed into a heavy snow-storm.

SEPTEMBER 23.—Deep snow lay on the ground this morning, which prevented the contemplated change of camp. The weather is bitter cold. The engineers, as well as nearly everyone else, have kept closely in camp.

SEPTEMBER 24.—More snow fell last night, and freezing weather continues. Camp was moved farther west into the mountains, about six miles. Crossing the divide we descended a range of steep hills, and camped in the snow upon a broad valley having a small stream, which, after many windings, finds its way into a branch of Deep Creek or Smith's River. The Fort Ellis and Camp Baker road runs through this valley, and we will take up our march on it in a day or two, the cavalry for Fort Ellis, and the infantry for Fort Shaw [to the north].

SEPTEMBER 25.—Camp Separation.—The infantry battalion left us today, taking the road to Fort Shaw via Camp Baker. The little stream on which we are camped is the most southern tributary of Smith's River. Two or three miles south runs a stream, heading in Crazy Mountains, which is the main tributary of Sixteen-Mile Creek, or Little Green River. Just across the ridge to the southeast is a branch of Shields River.

SEPTEMBER 26.—An early start was made this morning, and three companies of cavalry, which I accompanied, moved along the Fort Ellis road, leaving the other company with Colonel Haydon as escort for the surveying party. Our route lay along a nearly level valley, bounded on both sides by high hills, which gradually rise into mountain ranges. The road is nearly due south, and excellent, except where crossed by small rills, at which places there was some difficulty in proceeding on account of the mud. We crossed the small stream on which we were camped (viz. the south branch of Smith's River) and about four miles farther on crossed the principal tributary of Sixteen-Mile Creek; several smaller streams were also crossed, when after passing over a ridge the beautiful valley of Cottonwood Creek, a tributary of Shields River, was presented to view. This valley is extremely fine, lying entirely surrounded except toward the outlet of the stream toward the southeast by high snow-covered ranges of mountains; Bridger range to the west, Yellowstone Mountains to the south, and Crazy Mountains to the east. We encamped on Cottonwood Creek, eighteen miles from last camp.

[*No entries dated September 27 or 28.*]

SEPTEMBER 29.—Camp was broken about 6 o'clock this morning. Fort Ellis is forty miles distant. I had intended crossing the range through Flathead Pass, but the threatening weather prevented and I accompanied Colonel Baker through the Bridger Pass to the post. This pass runs nearly north and south, and is about twenty miles in length. In this pass are the sources of several of the tributaries of Shields River, which flow out at the north, also of Bracket's Creek, which breaks through to the east, about midway between the two ends of the pass.

SEPTEMBER 30.—I rode over to the encampment of a detachment of the Seventh Infantry, twenty-five miles distant from Fort Ellis, stationed upon Dry Creek, about six miles from the entrance to Flathead Pass. This detachment was placed here to guard the Gallatin Valley from incursions by Indians by way of the Flathead Pass, on the Blackburn, a few miles farther north. The country along the valley of Dry Creek has numerous farms already under cultivation, while along all the water-courses coming in from the mountains ranches are springing up in great numbers.

OCTOBER 1st.—I went up through the pass in company with Lieutenant Sanburn, who, with his detachment, was on that day relieved from that station by Company A of the Seventh Infantry. Six miles from camp we entered the pass. It has been extensively praised for the beauty of its scenery, and has also been mentioned as a practical route through the Bridger range for the Northern Pacific Railroad. We continued the ascent of the pass on horseback. The scenery along the trail to the summit, about three miles, is far more beautiful than anything I had observed during the entire summer, and reminded me of the gorges of the Upper Yellowstone country. The pass is narrow and heavily wooded with a luxuriant growth of mountain pine, affording a most refreshing shade, while Pass Creek, a cold, sparkling stream, added music to the otherwise perfect stillness of the gorge. Leaping and dancing from rock to rock the waters of this creek presented an ever-varying picture of extreme beauty. The ascent is, however, much too steep for a practicable railroad route, being probably three hundred feet to the mile. An old Indian trail leads to the summit, from which grand and extensive views were obtained to the east and west.

It is to be hoped that this germ of mountain scenery may be preserved as a pleasure-ground in its present natural beauty, it being so accessible to the inhabitants of the valley below.

Joseph W. Flenniken to Milnor Roberts, August 23

Dear Sir,

I send you herewith copies of Col. Haydon's report and letter.[35] I have made triplicate copies and mailed to Corinne [Utah] c/o Tim Henderson. I have held these hoping to get some further information from the Yellowstone party but as yet have not. From the day I left them, at the [Yellowstone] ferry until the day I received the news of the attack, I lived in a semi-fever—a dread that they would be caught "napping" and the commanding officer sleeping so sound that he would be hard to wake—I am afraid that my forebodings were too true. Military genius, skill and bravery amount to nothing when counteracted by intemperance.

I feared that Baker, would continue drinking—and he has. I felt so strongly that, while at Ellis I remarked to one of his friends with the hope that it would be repeated to Baker, that if he, Baker made a mistake—this time it could be fatal to him—cost him his commission—I do hope that a rigid investigation into this matter will be made and if it be true that Col. Baker has disqualified himself for the important and responsible position and duty assigned him, that he be made [to] feel the full weight of punishment which such conduct deserves. Too many human lives and too much public enterprise hinged on his personal conduct to allow him to go unharmed.

I write strongly on this subject—for I feel almost a personal responsibility myself—had I thought that speaking to Baker on the subject would have done any good I would have done it, but I feared that it might work in an opposite direction. I intend this for you personally and hope that you will not think me overstepping my duty in this writing.

Thomas' party got off yesterday.[36] . . . Mr. Burgess and his party are

---

[35]The report and letter have not survived. Joseph Walker Flenniken (1833–1907) was Helena's office manager, assistant engineer, and paymaster for the Northern Pacific. His father (President Polk's ambassador to Denmark) apparently had a business or family association with Cooke, for one of Flenniken's letters begins, "Dear Jay." Flenniken was a lawyer and district attorney before the war. He served for one and a half years as a sergeant in the Sixty-second Pennsylvania until being discharged for an unstated disability. The 1880 census shows that he was a clerk at his father-in-law's Pittsburgh savings bank. At the time of his death he was a vice president.

[36]Thomas P. Roberts (1843–1924) was Milnor Roberts's son. He had a distinguished career as a surveyor and after the Northern Pacific spent most of his life on waterway improvements in Pennsylvania and Ohio. Roberts was no stranger to adventure; during the Civil War, on his way to join his father in Brazil, his ship was sunk by a Confederate raider, and he was cast adrift, barely surviving.

at work. . . . While at Deer Lodge a few days ago I had conversation with Granville Stuart,[37] on a route via [——] River and Boulder River, I had concluded to make a B[——] trial survey myself when your order reached me for Thomas to make a survey. What caused the conversation was a trail that I wished Burgess pack train to take—it leads from Deer Lodge up GroFino[38] to the head of North Boulder, and direct from thence almost direct to Gallatin City. On inquiring about it at Deer Lodge several parties informed me that there was no such trail, but DeLacy[39] has it on his late map, and I have hunted up Mr. Stuart, who said that there was a good trail—and so the men found it—and saved about 75 miles—I have suggested to Thomas that when he reaches the head of Dog Run on his reconnaissance, he might return via GroFino to his party, whereby getting an idea of both routes without losing any time.

. . . I have a pretty tough time in keeping Mrs. Haydon's spirits up to zero—but Tom and I manage between us to convince her that the Col. is safe—should anything happen [to] the Col.—I don't know what I would do. When Tom gets off I will have <u>three</u> families to look after, rather a big undertaking for a young man, don't you think so—fortunately however Mrs. Roberts and Mrs. Haydon and my wife "hit it off" nicely—especially Mrs. R. and my wife—we are in fact one family—each helping one another and fully altogether. All of us send many kind regards to you and yours.

Very respectfully . . . [etc.]

---

[37]Granville Stuart (1834–1918) settled in Montana before the state's first gold rush. He was a leading pioneer, vigilante, and businessman, known as "Mr. Montana." There is considerable literature about him, including *As Big as the West* by Clyde A. Milner II and Carol A. O'Connor.

[38]GroFino was a creek. While it appears to be located in the Crazy Mountains between the Yellowstone and Missouri Rivers, I have been unable to locate it. Or has the name changed?

[39]Walter Washington DeLacy (1819–92) was a descendant of a Norman knight who fought with William the Conqueror. A few generations after that ancestor, another DeLacy was a significant participant at Runnymede, when King John was forced to sign the Magna Carta (1215). After college DeLacy taught language and mathematics courses at West Point and Annapolis, was an engineer with the U.S. armies invading Mexico, helped build the Mullan Road, participated in at least two Indian wars, was a pioneer civil engineer in Montana, assisted the early Yellowstone Park explorations, and worked on a consulting basis for the Northern Pacific. Today his name is found on one of Yellowstone Park's lakes.

Business Implosion

By late fall, various Northern Pacific troubles were springing leaks in the company's financial dike faster than Cooke could plug them. The lack of cash led to construction workers not being paid, their rioting (including the destruction of railroad property), and the railroad—in desperation—selling its inventory of unbuilt track. Most construction was halted, although some continued piecemeal.

There were numerous reasons for the lack of cash. Federally mandated land would not be approved until after Smith's exit, and thus sales from this source were zero. Freight and passenger traffic to Fargo-Moorhead was limited, although it would substantially increase in 1873. Funds had been directly and indirectly siphoned off through overspending. Also, unnecessary and ill conceived track had been built toward St. Vincent and Canada, partially to eliminate riverboat competition from Fargo-Moorhead and partially to preserve speculative land profits by Smith and others in Duluth. Nevertheless, the prime cause for lack of cash was Cooke's inability to sell Northern Pacific bonds.

St. Paul Pioneer, September 13

THE NORTHERN PACIFIC

CHANGE OF PRESIDENTS—

REASONS FOR GOVERNOR SMITH'S WITHDRAWAL.

FROM THE *NEW YORK EVENING POST*

The frequent rumors of the resignation of John Gregory Smith of the Presidency of the Northern Pacific Railroad have at last been confirmed by the actions of the corporation's board of directors. Mr. Smith's retirement from the office was induced partly by the condition of his health; principally by the demand upon all his time made by the Vermont Central Railroad combination, of which he is the head. The directors of the Northern Pacific have accepted the resignation of Mr. Smith, and elected as his successor Geo. W. Cass, distinguished in American railroad enterprises, and recently the President of the Pittsburgh, Fort Wayne, and Chicago line. Mr. Cass will take up his residence permanently in New York and give his entire time and undivided attention to the Northern Pacific. His name will give the assurance of a wise and successful management of the great enterprise he has been chosen to preside over.

FROM THE *PHILADELPHIA ENQUIRER*

Ill health having compelled John Gregory Smith Esq., to retire from the presidency of the Presidency of the Northern Pacific Railroad,[40] the directors of that great and thriving enterprise have chosen in his stead George W. Cass. A wiser choice than this could not have been made, Mr. Cass having deservedly achieved an enviable reputation as an administrator of various railroad enterprises, while performing the arduous duties of President of the Pittsburgh, Fort Wayne and Chicago road. The new head of the Northern Pacific is, therefore, a gentleman of ripe experience in railroad matters, keenly alive to all the needs of the road he has been called upon to direct, and is fully aware of the advantages which must accrue to the Northwest from the early completion of the Northern Pacific from Philadelphia to the shores of the Pacific ocean.

St. Paul Pioneer, October 4

Perhaps the most interesting and instructive point of the following article is that despite the Northern Pacific running out of cash in 1872, and despite office facilities in St. Paul and an existing, functional two-story wooden depot-office building in Brainerd completed in 1871, Gregory Smith still went ahead with the construction of this thoroughly unnecessary building.

[40] "Ill health" was similar to today's "exploring new business opportunities" or "spending more time with his family."

NORTHERN PACIFIC HEADQUARTERS AT BRAINERD

FROM THE *BRAINERD TRIBUNE*

The grandest in the New Northwest is the General Headquarters office building [and depot], now being finished off in Brainerd. It is a mammoth building, three stories high, besides the basement—which latter contains three fire proof vaults—with the heaviest, richest cornices and bracket-work, the whole surmounted by a Mansard roof, with heavy dormer windows and an immense tower rising from the basement away above the rood, ornamented at the top with an observatory and the richest finish. The whole building is surrounded by a great platform, and the offices inside for beauty, size, and general arrangement and fit-up will compare with any in the Western country. The whole structure is a thing we may all feel proud of, and all we Brainerdites are as proud of it as a little boy with a new knife.

New York Times, December 12

Federal approval was necessary in order for the Northern Pacific to sell some 2,880,000 acres (roughly 4,500 square miles) of government-owned lands in Minnesota allotted to it by Congress.[41] *The actual certification process was left to a three-man commission, its members appointed by President Grant. In fact, the commissioners were selected from a list of names submitted by the affable Henry D. Cooke (Jay Cooke's brother), governor of the District of Columbia and Grant's card-playing crony. To no one's surprise, the commission rapidly approved the Northern Pacific's work.*

[41]The 1864 Railway Act specified that the Northern Pacific would receive 12,600 acres per mile of certified road in existing states. Thus in Minnesota, when 228.8 miles were approved, this amounted to some 2,882,800 acres or 4,505 square miles. For territories such as Dakota, Montana, Idaho, and Washington, the grants were 25,200 acres per mile of certified road in 25-mile segments. One of the mysteries of Smith's presidency is why—with cash badly needed and land sales so obvious—he adamantly refused to file anything. Possible answers include that he financially benefitted or that initially construction was so shoddy that even "friendly" commissioners wouldn't have approved the shoddy construction.

THE NORTHERN PACIFIC

The Commission appointed to inspect the Minnesota Division of the Northern Pacific Railroad, prior to its acceptance by the Government, today made their report to the Secretary of the Interior substantially as follows:

"The section examined is found to be well located, both with reference to through and local traffic. The grades are light, three-fourths of the distance being below the average of thirty feet. The sharpest curve is three-degrees in a radius of two-thousand feet. The embankments and excavations are within the requirements of the law. The rolling-stock is uniformly of excellent quality and in good condition and is amply sufficient for all present and future requirements. The engine-houses and repair-shops are deemed to be adequate. The passenger and freight stations are specially commended for their neat and tasteful appearance. Mention is made of the emigrant-houses at Brainerd, Glyndon,[42] and Duluth, as being constructed with the furniture and cooking utensils on the Company's or the Government's lands. The snow fences, water stations, and telegraph lines are considered ample for their purposes."

The report closes as follows: "Having found this portion of the road to be judiciously located, well constructed, and adequately equipped, and believing that it substantially meets the requirements of both the letter and spirit of the Law and Department regulations, the Commission recommends its acceptance by the Government."

The report covers a distance of 228 miles.[43]

Philadelphia Inquirer, December 12

THE NORTHERN PACIFIC RAILROAD

The following is a full synopsis of the report of the commissioners appointed by the President to inspect the Minnesota division of the Northern Pacific Railroad. The report is signed by all the commissioners, viz., S. H. Kaufman, A. C. Stands, [and] Thomas Underwood.[44]

---

[42]Some ten miles east of Moorhead.

[43]The detailed report lists the commission-approved distance as 228 miles—4,118 feet from Thomson Junction, some 20 miles southwest of Duluth.

[44]Sections discussed in the *New York Times* article are omitted.

... BALLASTING. The road was found to be well and substantially ballasted with gravel. The ties are firmly embedded, and the track is smooth and in good condition. In illustration of this point, it may be proper to say that regular passenger and freight trains have been running at fully average rates of speed over the entire length of the division since the first of last May.[45]

... CROSS TIES. The ties are laid twenty-six hundred and forty to the mile . . . and freely conform to the prescribed standard.

... RAILS. The rails are of American manufacture,[46] and weigh fifty-six pounds to the yard . . . [and] meet all the requirements of the government on this issue.

... ROLLING STOCK. The rolling stock of the company is constituted as follows:

Locomotive engines[47]	68	[Not including 28 ordered and under construction.]
Passenger cars	18	[Not including 4 under construction.]
Baggage cars	5	
Pay cars	1	
Observation cars	2	
Box cars	350	
Platform cars	1100	
Caboose cars	6	
Wrecking cars	1	
Hand cars	52	
Rubble cars	36	
Pile Driver	1	

[45]It is not clear if "last May" means 1871 or 1872. In either case, the paperwork to begin the approval process was running a half year behind schedule, a direct swipe at Smith's management.

[46]Congress had slipped into the Railroad Act a requirement that the iron be American made. This provision also diminished Northern Pacific's prospects for European funding.

[47]What did the Northern Pacific pay for its locomotives? Any answer must be subjective, but given the variables in locomotive manufacture, in the early 1870s it was a seller's market. The Northern Pacific's supplier, Baldwin Locomotive, was the nation's leading builder with a reputation for high quality, state-of-the-American-art technology, albeit lesser quality than more expensive British engines. In general, heavier locomotives of the type the Northern Pacific ordered cost more, but this should have been offset by their large orders. My estimate is that costs per engine (and tender) were near the top of the day's spectrum, in the $11,000 range ±$1,500, but not including a 5 percent war tax *(continued)*

. . . PASSENGER, FREIGHT AND WATER STATIONS. There are fifteen passenger and nine freight stations. . . . The water stations are twenty in number.[48]

Perhaps the most striking elements in the above are the excessive number of locomotives and the comparatively small number of passenger cars. As for the lone pile driver, given Minnesota's constant construction problems through over a hundred miles of wetlands, that number didn't increase and was all but criminally negligent. Additionally, it is known that in 1870 the Northern Pacific owned five large steam shovels, ideal equipment for Dakota and the land west of Minnesota's lakes, bogs, and swamps. Where these had gone by 1872 is not known.

Helena Daily Herald, November 12

IN TROUBLE

No one will question the friendly attitude of this journal towards the Northern Pacific Railroad, or the consistency with which it has advocated its construction across the continent. No paper in the whole land has given up so much of its space in efforts to encourage the commencement and afterward to promote the speedy building of the great trunk line into, through, and beyond Montana, to its designated western terminus on the Pacific coast.

We have time and again referred to the many advantages which this northern railway line would possess over other completed or contemplated roads connecting the two oceans, when finished. We are no less a Northern Pacific enthusiast than before, though obstacles intervene to prevent its building as rapidly as we had hoped. Our Territory suffers immensely for lack of railroad communication—for want of sure and quick transit—connecting this northern interior with the great markets and population communities east, west, and south of us.

---

likely still in effect. A higher cost would also reflect Gregory Smith's management, for if there was some way to get "free" locomotives for his Vermont Central, he would have done so—and at the Northern Pacific's expense. If these ballpark figures are accurate, the Northern Pacific would have paid almost $750,000 before they ran out of cash. See John K. Brown, *The Baldwin Locomotive Works, 1831–1915*, and John H. White Jr., *A History of the American Locomotive: Its Development, 1830–1880*.

48. A twenty-first water station was fed by a stream or small falls and counted separately.

There is trouble apparent, which threatens delay in the coming of the Northern Pacific, and while we sincerely regret this, and would contribute all in our power to obviate the difficulties which harass the Company in their efforts to master the situation, we must look beyond and in another direction for a solution to our sorely felt need for speedy railroad [service] to and from the country.

The North and south line is our current hope,[49] and its achievement is possible, nay *certain*, if our people can be brought to realize the existing emergency for action, and come forward with prompt assistance to the enterprise. We append hereto an extract from a Minneapolis, Minn. Banker (formerly a member of this city), which extract gives an unquestionably correct statement of Northern Pacific affairs at this time. Let our people ponder this matter and say then if there is reason for any person in Montana to interpose further objections to securing the North and South railroad.

"The Northern Pacific Railroad Company owes, in this country, about six-hundred thousand dollars, and it seems that there is no funds with which to pay up.[50] It is creating almost a financial crisis here and at St. Paul. It is hoped relief will soon come. All the works have been suspended for thirty days in consequence. The track is laid within thirty-five miles of the Missouri River."[51]

Helena Daily Herald, November 30

NORTHERN PACIFIC RAILROAD AFFAIRS

WE COPY THE FOLLOWING FROM THE *CINCINNATI COMMERCIAL.*

It is known in railroad and financial circles that the St. Paul and Pacific branch of the Northern Pacific has ceased work for the want of funds and the orders for cars are countermanded. This sudden halt is said to be owing to the caution of Mynheer of Amsterdam,[52] who refuses to make

[49]Corrine, Utah, was a transcontinental railroad stop due north of Salt Lake City.

[50]Although unclear if the $600,000 referred to Montana or to Montana, Dakota, and Minnesota, my guess would be the latter.

[51]The *Herald* does not state the source of this quotation.

[52]The lead Dutch firm was Kerkhoven & Company of Amsterdam. The name "Mynheer," whether an individual or a company, is nowhere to be found in Augustus J. Veenedaal Jr.'s *The Saint Paul & Pacific Railroad* (1999). *Mynheer* is the Dutch equivalent of *Mister.*

further advances on the bonds of the company. He was assured when the road reached a certain point it would pay expenses and interest; instead of doing this, the road has from January to July of this year fallen some forty thousand dollars behind expenses, to say nothing about interest. Not an unnatural caution. Now, as the branch is estimated to be worth, mile for mile, twice as much as the main line, the outlook is not assuring. It is understood the Northern Pacific managers are getting ready for an appeal to Congress for a money bonus in addition to their land donation.

ALSO THIS FROM THE *OMAHA HERALD*

Information has been received from Minnesota saying that the work of the Northern Pacific Railroad has been suspended, that the contractors had left, and that the laborers and [mule and horse] teams had returned from the end of the track at Ross Lake, Dakota, thus leaving 30 miles of the road to the Missouri River unfinished. Failure to sell bonds and want of money seem to be the trouble.[53]

Messrs. Jay Cooke & Co., the financial agents of the [northern Pacific], are of the highest financial standing in the monetary world, and their efforts have been almost superhuman. We trust there is some mistake about these reports, and that the work is yet progressing, but private advices from Minnesota confirm the statement that there is a temporary embarrassment in the affairs of the company. Montana must make her influence felt in aid of this matter. The Territory should publish a reliable, complete statement of the resources which assure to the Northern Pacific Railroad a future of wealth such as few railways in the world possess.

[*The remainder of this article appears to have been written by the* Helena Herald.]

We do not give full credence to the newspaper extracts, but we must have the North and South railroad, for its construction will be a most welcome help to the Northern Pacific. Mr. Cass, the new President of the road, is a clear-headed, far-sighted railroad manipulator, and we look to him, the directory, and the financial agents to retrieve the situation of the company.

---

[53]The other causes included unnecessary construction toward Canada that duplicated Northern Pacific track, lack of construction on the strategic St. Cloud to Brainerd route, and ongoing corruption.

PART IV

Last Reports

While the 1872 surveys ended in mid-October, there were still reports to write. Stanley's final section is a detailed botanical and geological description of the land, not coincidentally subjects in which he had excelled at West Point. His conclusions also offer a highly critical analysis of army operations as he touches on inadequate grain for horses, poor ammunition, "flimsy" tents, pressed paper replacing leather in shoes, too little bread ("the complaint of hunger was constant"), a broken axle that was painted over, and too many physically "feeble men." This section especially deserves reading to understand the difficulties that he, or any commanding officer, might have experienced.

In sharp contrast, Baker's entire report is as useless as it is short. Literally under army arrest and knowing that he might soon be charged, he was careful to say as little as possible, his description of the fighting limited to one sentence. As to his return, Baker falsely wrote, "After the Indians had retired the camp was moved and the Survey continued *as usual*" [emphasis added].

No one better described the Northern Pacific's problems than Montgomery Meigs. On October 1 he noted that the nearest locomotive was 80 miles from Bismarck, and he accurately predicted that the surveyors would be laid off on their return, although he was retained until year's end. Meigs's last letter on December 27 mournfully describes his trek for three weeks and 350 miles by sled paralleling NP track across snow-covered prairie.

As for the unfairly disgraced western surveyor John Haydon, he and his young wife disappeared without a trace.

Stanley's Official Report: "Observations"

At first glance nothing might seem duller than a section entitled "Observations." However, when Stanley angrily begins to describe the dishonesty of army contractors and the ineptitude and laziness of the army's own quartermaster, ordinance, and medical departments, this section proves to be fascinating.

Having finished my military report, I will add some description of the country, and some remarks on other subjects, that may be of interest or use to the Commander of the Department

Description of Country

The region lying due west of Fort Rice, and as far as the mouth of Powder River, embraces several varieties of soil, some difference of climate, and quite a difference of grasses and plants. The first belt of country of like products extends from Fort Rice to second crossing of Heart River, marked Sully's Crossing. This country has a rich loamy soil, producing a very luxuriant crop of grass (mixed Buffalo and Prairie grass), and for a prairie country it is exceedingly well watered. It only needs security of life, and protection from prairie fires, to convert this country into a great stock raising region. I believe the small grains, oats, wheat, and barley, will succeed in this country. The surface may be described as gently rolling prairie, breaking down by sharp hills in the vicinity of the streams, the scenery lightened up by the occurrence of these low flat topped mountains, called "Buttes" by the French, and which almost invariably indicate the origin or course of the streams which drain the country. The tops of these buttes indicate apparently the original surface of the country, and their present form seems that small part of the surface which from its hardness or its position has resisted the action of water through countless ages. From Rice west the first days travel you have Little Heart Butte on the right toward Fort McKean, the second day the Three Buttes appear 10 miles to the left of the road and near the Cannon Ball River. The Dog Teeth Buttes are peculiar in form (serrated). They are forty miles west of Fort Rice, and immediately upon the road. From these latter buttes, the true Heart Butte is seen; a single cone slightly flattened at its apex, twenty-five miles to the west, and five miles north of Heart River, a land mark visible for many miles in all directions.

The Heart River, which is mainly followed by all the routes west from Fort Rice (excepting Sully's route on Cannon Ball), is a stream upwards of 180 miles in length rising by several sources in the "Bad Lands" or rather the rim of the "Bad Lands" of the Little Missouri. So far down as the Heart Butte the valley is narrow, not exceeding one third of a mile in width, from this point the valley spreads out to three fourths of a mile in width, and is apparently very fertile. The stream has only a fringe of timber, ash and elm upon its upper course, and cottonwood mostly from Heart Butte to its mouth. The past summer was perhaps exceptional in abundant rain, and Heart River was perhaps unusually full. When we left it, the stream on the shallows was twenty feet wide, and from six inches to one foot in depth, of excellent water. For camp and grazing purposes this stream has always abundant water and wood from its source to its mouth. The description of Heart River as to wood, permanency of water, and source, applies to Cannon Ball, but the banks of the latter and the hills enclosing its valley are not so abrupt as the Heart River. The bed of Heart River is about 130 feet wide, the banks averaging thirty feet in height, and very steep. The stream has the peculiarity of cutting nearly straight across its valley, making a carom upon the hill upon one side, and rebounding as it were to the other side, in each case making a precipitous bluff bank where it strikes the hill, in consequence of which a road in the valley of the Heart River is an impossibility, and a railroad can only be built by continuous bridging, as often as three bridges in one mile in several places. The region I have described up to 100 miles west of Fort Rice abounds in game; antelope are very abundant, and there would be no difficulty in the way of Indians feeding themselves abundantly upon game. Elk, black-tailed deer, and wild fowl are plenty.

From Sully's crossing to the Bad Lands of the Little Missouri, the soil becomes poor and very hard, having a burnt appearance, and a very poor covering of buffalo grass; the cactus appearing abundantly.

The two beaten routes, going and returning, are south of the major Heart River, the returning route keeping mostly on the divide between the latter river and the Cannon Ball. Springs are abundant, and wood is sufficient for camping purposes.

Immediately after leaving the sources of Heart River, up to which points the surface has been a gradual ascent, we come upon the "Bad Lands" of the Little Missouri, which from a well defined rim fall down some five hundred feet, in from five to ten miles, as the river happens to bend toward, or from, the general slope. It is very hard to give any adequate idea of these "Bad Lands" by writing a seemingly confused collection of bare hills, generally of the shape of a cone, or some modification of the ellipsoid separated by deep ditches amounting to charms, grassless, treeless and lifeless, varying in color from a pale clay, to a quite bright red, silent and solitary, these "Bad Lands" are indeed curious features of nature. They seem to be formed by water, it appearing plainly that at some period the rapid subsidence of great waters have stripped everything that could ever become soil, and left the bare clay, which rain in time has worn into these fantastic hills.

As we get near the Little Missouri large flats occur covered with good grass, and all the ravines that lead down to the river are well wooded with pine and ash. The Little Missouri itself has a bed 200 feet wide and, on the several occasions upon which I saw it, was a bold running stream of sweet water. The valley is quite narrow not exceeding one half a mile in width. The clay hills abound in fossils, the most notable of which are the gigantic ferns (I suppose them to be) varying from the size of a man's thigh, to specimens 300 feet long and ten feet or more in diameter. Coal crops out everywhere, and beds were found exposing a thickness of 15 feet. I did not find any coal on this stream which did not appear to be imperfect, only a little better than lignite. Should it ever become advisable, a military post could be built and maintained on the Little Missouri. Pine, though not in great forests, is yet sufficient for the construction of sample buildings, and both wood and coal are abundant. The grazing near the river is excellent and the sites for a post very fine. The only deficiency is hay which would have to be drawn from the ravines some distance off. I speaks particularly of this point 170 miles west of Fort Rice, as a probable and favorable place for a post, and will add that it is the only point on my summer's route where a post could be maintained, but I do not recommend it for a post, as it is not central in relation to the hostile Sioux, and could do but little good. Upon this stream we killed grizzly bear, mountain sheep, elk, black-tailed deer, and antelope. It does not seem to have been hunted much, and is now a splendid hunting ground.

The Red Butte marks the passage of the road into the "Bad Lands" on the east. From this point the Rainy Butte is seen 15 miles distant east; the great white Buttes, s. of S.E. 20 miles, the cow Buttes, 30 miles south; The Buttes which look at each other on the Little Missouri, 10 miles due west. Going west from the Little Missouri the soil becomes poorer and the grass more sparse. After passing Beaver [Creek] which presents the curious feature of a creek upon a ridge, we fall upon the "Bad Land" of Cabin Creek which only differs from those of the Little Missouri in not being so wide. The valley of this creek where we struck it is four miles wide contracting to one half a mile in width before it reaches the Yellowstone; it has a growth of cottonwood on its banks, in numerous places large groves of heavy timber. The valley is poor and mostly grown over with sage and cactus. The hills have numerous cedar brakes, grass very poor and scarce. Here and upon Beaver [Creek] we found abundance of sage hens, beautiful and excellent game. From Cabin Creek west to O'Fallon's Creek the same high poor prairie, cactus, sage, short buffalo grass, no game, plenty of rattlesnakes.

We struck O'Fallon's Creek (Indian name, "The creek that makes a hole in the rock") 26 miles from the Yellowstone. This creek is a stream of fresh running water, has a large water bed, 200 feet wide, has a growth of cottonwood at times quite heavy, all the way to its mouth, the valley is three-fourths of a mile wide and the hills have the same bad land appearance of the Yellowstone and Cabin Creek.

The stream itself of the Yellowstone for the thirty miles or more that I observed it is a beautiful river. It is apparently 200 yards wide, deep, rapid, and in August, clear. When we have praised the river we are done. The valley, from one to four miles in width, is poor, very sterile, the grass very indifferent, abundance of sage and cactus (always inseparable from poverty), and the timber upon the Yellowstone and Powder River so far as I saw would not build one decent cabin. Hard coal is abundant and apparently very good. This year the river kept up all summer from the abundant snow, and was nowhere fordable. The rapid below Powder River [Wolf Rapids], I should think no obstruction when the river is at all full, indeed by taking out a line, and a boat could have passed it drawing three feet when we were there.

I saw no place upon the Yellowstone suitable for a military post, but a post could be maintained at the mouth of O'Fallon's Creek or 20 miles below at the mouth of Cabin Creek, as the wood on either of these creeks would build a military post, and furnish fuel for years to come. No building stone was found on the Yellowstone, and hay would have to be shipped probably from a long distance. The guides describe a fine country west at Tongue River, and the Rosebud, and my impression is that the site for the future post must be searched for at the mouth of one of these streams. I am decidedly of the opinion that a large post must be established upon the Yellowstone, a post strong enough to crush any band of hostile Sioux, before the murdering and plundering bands can be made to respect the Government. The post should be on the Yellowstone itself with a steam ferry. The hostile Sioux cross and recross the Yellowstone at their pleasure, and we must make either side unsafe for them. The completion of the Northern Pacific Railroad would make all this an easy and plain matter. If the road is going to be delayed several years, an attempt to supply a post by steamboat during the high water of next spring should be made.

The country all the way from the head water of Heart River to the Yellowstone retains traces of immense herds of buffalo, which seem from the sign to have nearly all disappeared eight or twelve years ago. Only a few buffalo were seen on the trip. The valley of O'Fallon and Cabin creeks have been great wintering places for Sioux, and immense quantities of dead cottonwood trunks lie upon the branches during winter. Why the buffalo should have all left, or why they should have ever ranged there, the country being so poor, I cannot conjecture. Finally the climate of the Yellowstone country was noticeably cooler than that of the Missouri and everything indicated that the rain fall was very light during the year.

The practicability of running a railroad up the valley of the Yellowstone without crossing and re-crossing is doubtful. In our observation of a very short part of the river, we found making several precipitous cut bluffs, where it would cost less to bridge the river than to cut a railroad through the bluff.

Quartermaster's Department

TRANSPORTATION

The Government train sent to Fort Rice from Abercrombie was well organized and efficient, but was not properly fitted for a march over the prairies. The train was lacking in extra parts for wagons; deficient of Lariat rope and hobbles. This deficiency was made up as possible by taking parts from the new wagons (12 in number) at Fort Rice. Several of these new wagons were taken entire, and broke down after a few days use. The wagons now furnished by the Quartermaster's Department are brightly painted and look new, but the wood work is dry rotted, and in two cases we found the iron of the axle broken but painted over. The contractor's two-horse train from Minnesota was the finest I ever saw. The entire outfit was bright and new. If the horses could have been fed their half bushel of oats each daily, there is no limits to the amount of work such a train could do. But we could only feed five pounds of oats, and part of the time no grain at all. Under these circumstances the train failed, and had I [not] sent the Minnesota train back to bring up supplies from Fort Rice . . . the train would have wholly broken down. One pair of little Indian mules will do more work upon the plains in one summer than the finest pair of horses, when each comes down to grass.

CAMP AND GARRISON EQUIPAGE

The tents now furnished troops are very flimsy, and one summer's service wears them out. The companies of the 17th Infantry, and the Indian Scouts started out with worn tents, and before the end of the march were almost entirely without shelter. If tents are carried at all they should be new upon the start.

CLOTHING

Each man started with three pairs of new shoes, and though this was one pair of shoes to the month, and several hundreds of pairs were issued, the command returned in a very bad condition as regards shoes. The shoe now issued is an imposition; some that we examined had pressed paper for insoles and middle soles.

Subsistence Department

I have only to remark of this Department that they were not on time with either bread or meat. There is a great waste in hauling pork over long distances as we had to do when we sent back for supplies. The ration of hard bread, one pound per day is not enough, men marching all day in the stimulating climate of the plains, need more bread, and one pound and a quarter is not one grain too much. The complaint of hunger was constant, and company commanders, as well as private soldiers individually, came to me every day begging the favor of being allowed to purchase bread. This I was obliged to refuse. But the question comes up in my mind, whether, in a country like our own, where provisions are so cheap, it is sensible, politic, or humane to send soldiers to the field, to work hard, day after day, feeling the craving of hunger. The Engineers feed their men one pound and a half of flour daily and their labor in not as severe as the soldiers. Surely, officers of the Subsistence Department, who fix the ration, are not competent judges of the wants of the soldiers, disconnected as they are entirely from military life.

Ordnance Department

The two banded breech-loading rifled musket seems all that can be desired in a serviceable gun. The men all carry the belt of their own make, with the cartridges arranged about the body, inserted into thimbles or pockets constructed in the belt. This appears the common sense plan of carrying the present ammunition.

The Spencer's Carbine ammunition is fully one half of it bad, and will not explode; it should all be condemned and replaced by good ammunition [emphasis added].

I carried three Gatling guns but am not in favor of them for field service. The small Gatlings are not superior in range to the breech-loading musket, and the large gun though of very long range, is so heavy that it requires six mules to draw it, and is not a gun equal to the three- inch Rodman, for effect, either in killing or scaring Indians. It seems to me that it argues a great want of enterprise, or great ignorance, upon the part of our Ordinance Department, that we have never been furnished a shell gun equal to the French 4-pounder or the German 6-pounder for Plains service. As our Ordnance stands now I should ask for three-inch Rodman guns, were I to command another Expedition on the Plains.[1]

The side line hobbles now made by the Ordnance Department is the only absolute security against stampeding animals whilst grazing. I recommend that every post in this District be supplied with them and ordered to use them. I tried to get side-lines before starting out but found that a protracted correspondence with an Ordnance Officer was the preliminary to getting anything, and had not the time.

As we are to have a change of knapsack, I will pass over the present disgraceful ones without comment.

Medical Department

Upon the medical Department I desire to say that through the fault of either the Medical officers of the Army, the Captains of companies, or commanding officers, too many broken down or worthless and feeble men are kept in service. This was noticeable in all the regiments, but notably in the 8th Infantry. After leaving sick men at Randall, Sully and Rice, we sent back twelve sick men from the Little Missouri and yet a

[1]This is exactly what happened in 1873.

few days afterward a sick list of 16 men appeared and one day reached 24 out of an aggregate of 266 enlisted. I think Surgeons and perhaps Commanding Officers are in the habit of saying, when a case is brought up: "Oh! he is a good garrison soldier, keep him," without reflecting what an embarrassment, a nuisance indeed, such a soldier is to a commanding officer in the field. Unless a soldier is good for field service, he is good for nothing, and the sooner discharged the better.

I intended to make some remarks upon Indian matters in this report, but will try and write upon the matter in a separate paper.[2]

I am [etc.]

Maj. Eugene M. Baker's Official Report

Fort Ellis M. T.,
October 18, 1872
Assistant Adjutant General, Department of Dakota, Saint Paul, Minn.
(Through Headquarters District of Montana)

Sir:

I have the honor to make the following report of the operations of the Escort to the Surveyors of the N.P.R.R. Company under my command.

The Escort consisted of four companies of the 2nd Cavalry from Fort Ellis and four Companies of the 7th Infantry from Fort Shaw. Total number of commissioned officers Twenty (20)—Total number of enlisted men—three hundred and forty-nine (349). Major J. W. Barlow, Corps of Engineers and 1st Lieut. J. W. Jacobs R.G.M. 7th Infantry accompanied the Expedition, and are included in the total commissioned. There were also about forty-five (45) teamsters and employees and citizens all armed. The Infantry portion of the command left the Post on the 27th of July, the remaining portion following on the 28th. The command was consolidated on the Yellowstone at the mouth of Shield River about 35 miles from the Post. From this point the command and Surveyor moved down the Yellowstone to a point about seven (7) miles below the terminus of last year's survey, keeping on the north side of the river the entire distance, the depth of the stream [Yellowstone] rendering it impassable.

This part of the journey was made without incident and we arrived at the above mentioned point on the 12th of August. We remained in

[2]I have been unable to find this separate report; perhaps it was never written.

camp on the 13th. On the morning of the 13th the Surveyors returned to the point where the Survey of last year terminated and brought their operations to the vicinity of camp. About 3 A.M. of the 14th the command was attacked by a band of Sioux and Cheyenne variously estimated from 400 to 1,000. The companies were promptly formed by their company commanders and the Indians were easily repulsed. Our loss in the engagement was as follows. One Sergeant of "C" Company 7th Infantry killed, and two Cavalrymen and one Infantryman wounded. One citizen was also wounded and afterwards died. 15 beef cattle and 4 mules were driven off by the Indians, and two cavalry horses died from the effects of wounds. The bodies of two Indians were found on the field and fourteen (14) dead ponies.

After the Indians had retired the camp was moved and the Survey continued as usual. On the evening of the 19th of August Mr. Haydon, Engineer in charge of the party, sent me a note saying that it was, in his opinion, impossible to proceed with the Survey, as it would be necessary to divide the command in order to get the transportation through the bad lands, and he did not consider the divided forces sufficient to conduct the transportation and escort his party safely, and therefore declined to continue the Survey on the Yellowstone, and requested me to conduct his party over to the Musselshell river, in order that he might prosecute a Survey up that stream, or if my order did not permit of such disposition, to conduct the party back to Fort Ellis.

In compliance with this request I broke camp on the morning of the 20th of August and marched over to the Musselshell. From this point the Survey was continued up the river to little deep creek on the Fort Ellis and camp Baker road. At this point I broke up the command, sending the Infantry to Fort Shaw, via Camp Baker and returning to Fort Ellis myself with three companies of Cavalry, leaving Company "L" 2nd Cavalry with the Surveyors until their services were no longer required, or until relieved by proper authority. A map of the route will be forwarded so soon as completed.

I am [etc.]

Source: Maj. Eugene Mortimer Baker, "Report of the Operations of the Escort to Surveyors of the N.P.R.R. Company Down the Yellowstone," retyped copy, Montana Historical Society, Helena. No paragraphs in original.

Letters of Montgomery Meigs

Fargo, October 1

We are having lovely weather though old Boreas gave us a foretaste of what is in preparation for us next winter, a few days since. I drove forty miles in an open wagon in the teeth of the storm of rain and sometimes a little snow with fifty miles an hour wind.

I have a nice office now. It is nearly finished. The walls are double and the space between filled with sawdust. I have a telegraph office and a powerful battery arranged on shelves high up in rows. The operator, Collins is a pleasant companion and has taught me [to] telegraph quite well.[3] I send nearly all my own telegrams now and have a great many. We use the telegraph a great deal and often Col. Gaw, the acting Div. Engineer, comes into his office at Fargo, two hundred miles away, and asks questions which I hear immediately at Edwinton. The telegraph is certainly the most wonderful invention of the age.

I went down to the end of track a few days ago and saw the locomotives now about eighty miles from here [west of Jamestown]. I heard a good deal of news, among other things that the report was in circulation that all the engineers were to be discharged this winter as the company did not intend making any more winter surveys. This will suit me well enough.

The new president, Mr. Cass is making a clean sweep, cutting down pay and discharging employees.[4] I think that the [Smith] faction is about played out on this road and the Jay Cooke [faction] is going to run it to suit themselves. There seems to be a sort of crisis in the affairs of the RR and there are rumors as there were last year about this time that the road will go no further than the Missouri for some time to come.

---

[3]Either William or Wallace Collins, brothers, both surveyors.

[4]George Washington Cass, 1810–88, West Point '32, was assigned to the Corps of Engineers. Nephew of Gen. Lewis Cass, who was governor of the Michigan Territory, a U.S. senator, secretary of state, and the 1848 Democratic presidential candidate. Andrew Jackson appointed Cass to lead the construction of the "National Road." He remained with this project until its completion through Maryland, Pennsylvania, and (West) Virginia. Afterward Cass was employed in stage and canal transportation, but mostly railroads. Cass was an early leader in the freight business. Insightful, honest, and modest, he joined the NP's board of directors (ca. 1867), became its president in 1872, and kept the railroad operational—despite legal bankruptcy—through the worst of the Panic of 1873.

Bismarck, October 19

I fear the [Northern Pacific] is in a tight fix. It is said that money does not come in rapidly enough now to carry on the work and we have not yet been paid off for August yet. Gen. Rosser telegraphs me however that the money has been received and will be forwarded at once. The company is [in tight circumstances] but I have no real fears for its financial stability. Part of the [eastern] Yellowstone expedition has returned all in good health and spirits. Half has gone back for a three weeks trip up the Heart River. I fear that the line of this year is a failure. The profile is simply impracticable. Gen. Rosser is of the opinion that the line of last year will have to be built on. . . .

A perfect Indian summer. Gen. Rosser offers me employment this winter. The Engineer Corps generally is to be discharged. Everything here will be closed up; no trains will run till the spring. They are anxious to reduce expenditures evidently.

Fargo, December 27

I have been out to the end of the track again and returned, a trip in sleighs of three hundred and fifty miles. I had two teams, a rodman, cook and teamsters and plenty of blankets and provisions but it was as you may suppose a cool trip. We slept at night at the little abandoned telegraph stations along the road and pitched tents for the horses. We had many days when the thermometer was twenty degrees below zero and the wind blowing a gale at that. One night we spread our blankets in a box car very open to the winds but as the thermometer was not much below zero we thought it quite mild and went to bed very comfortably. Next morning when we woke the wind was blowing hard and the thermometer read twenty-six below zero.

We are still having a continuance of the cold snap and for over a week the mercury has not been above zero. Generally the mercury has been frozen up in the morning and at ten o'clock the other day the thermometer standing in the full sunlight in a warm place marked twenty-six below. A day or two ago at eight o'clock three different thermometers indicated fifty-one below. What do you think of that? Strange as it must seem to you it was quite possible on that morning to stand bare-headed and handed out of doors and no one seemed aware that it was particularly cold. I think one suffers much less out here provided he keeps indoors when it blows.

We are now getting up a final estimate of the work done on this division and expect to get through by the last of the month. We are all to be paid off then and I shall be out of a job and off for home. I expect to take a small hunt on the way however over in Wisconsin and hope to kill a moose to balance Father's buffalo and bring the hide home with me.

The NPRR seems to be hard up. They pay all employees in full but the money for their contractors comes in but slowly. They have so disbanded the fine engineer corps we had and were so proud of that I scarcely care whether I stay or go. I feel though that I ought to make all the money I can now as it is improbable that I could get another job this winter and if times are dull perhaps not even next spring. None of the engineers who have left the road have obtained permanent employment so I ought to be well satisfied at being kept so long.

Aftermath and Conclusions

The Northern Pacific's report to the Department of the Interior (see epigraph) complained that the surveying parties had been stopped by "superior force." But were Sitting Bull's warriors really militarily superior? In fact, in tactical terms, they were defeated both times they gave battle. However, Baker's not crossing the Yellowstone turned out to be the key to 1872's failure. Had Baker and Stanley met, and even if Adair had died, 1872's primary goal would still have been obtained, albeit with painful memories. However, Sitting Bull's force of character proved too much for the self-confident, out-of-control drinking Baker.

Nevertheless, one of 1872's greatest ironies is that given Baker's documented weaknesses, this writer strongly believes that *if* he had crossed the Yellowstone at Livingston or Big Timber, he undoubtedly would have been subject to a reasonably well executed attack or ambush by two to three times the number of warriors he faced August 14. The result? Depending on the actual geographic location and always mercurial Indian *discipline*, losses the magnitude of Fetterman's or even Little Bighorn might easily have occurred.

* * *

In St. Paul Hancock was seething; his October 3, 1872, annual report hinted at what was to come for Baker: "It must be admitted that, in not proceeding to the mouth of the Powder River, part of the objects of this expedition will not be accomplished this year. Whether the force was inadequate, the action of the commander judicious, or whether the responsibility for not proceeding farther rests with the railroad engineers cannot be determined until we have further advices. [Col. Gibbon] has been directed to make an investigation."[1]

Hancock placed Baker "in arrest" October 15 and likely would have prosecuted him, but for the November 6 death of Maj. Gen. George G. Meade. This created a Sheridan-level vacancy in the east which became Hancock's by right of seniority. Hancock's successor, Alfred Terry, dropped the charges against Baker, but stripped him of his Fort Ellis command and had him purchase horses in Nebraska. Later, Sheridan gave Baker better commands, but his future was behind him and he died of alcoholism in 1884.

Major Barlow's career was damaged by his report, and Sheridan transferred him to the Corps of Engineers, where he languished for years. In 1901 Barlow was promoted to brigadier general, commander of the Corps of Engineers a single day before his retirement.

* * *

Meanwhile the army, infuriated and eager to even the score, began planning for 1873. That force equaled the combined size of all previous '71 and '72 surveys, more than 1,550 officers and men, included Frederick Dent Grant, who insisted on going to avenge his late cousin, Lieutenant Adair. One thing that Sheridan and Julia Dent Grant both wanted was that her son would be well protected. With young Grant as leverage, Sheridan assigned ten of the Seventh Cavalry's twelve companies to 1873's survey led by the Seventh's lieutenant colonel, George Custer. Col. Samuel Sturgis, who by rank should have been assigned this command, was banished by Sheridan to desk duty at Fort Snelling, Minnesota. Only after Custer's death (as well as the death of Sturgis's son at the Little Bighorn) was Sturgis again given field command, only to be brushed aside and humiliated by Chief Joseph in 1877.

Poor Stanley. This woefully unhappy man, having performed in an entirely credible manner, became the army's compromise choice to lead 1873's survey. But Stanley proved a deer caught in the headlights of Custer's aggressive persona and became submissive within a month after the survey began. Likely embarrassed by his placing Custer "in arrest" and a three or four day drinking binge, Stanley allowed Custer to write

---

[1]General Hancock's annual report to Sheridan, October 3, 1872. Hancock's remarks appear on p. 40 of the Secretary of War's Annual Report to Congress for 1872, published in 1873.

the report of the August 4 and 11 fighting. In this report Custer never mentioned his own name or Stanley's, but made it clear that without *his* leadership the resolute Sioux might have prevailed. Custer's gem, dashed off at high speed and marvelous in its descriptive powers, reached the *New York Tribune* in early September. Widely reprinted, it immediately restored Custer's national reputation.

Meanwhile Stanley's drinking had soured the survey's "embedded" correspondents. If 1873 etiquette forbade even hinting at Stanley's alcoholism, the writers simply stopped mentioning him, focusing instead on Custer and, to a lesser degree, Rosser.[2] On the expedition's last night in September, Custer had his band serenade Stanley. But by then Stanley's slide into obscurity had already begun.

As apparently promised by Sheridan in 1873, Stanley received an eastern posting the following year. He then assisted Sheridan through the Byzantine army political infighting of the late 1870s—getting court-martialed for his efforts—and, after Sherman's retirement, was appointed a brigadier general.

The 1872 surveys ended any potential Northern Pacific railroad domination in northern tier states and territories from Minnesota to Washington. Cooke and the Northern Pacific losing control of two railroads, the St. Paul & Pacific (later the keystone of James J. Hill's Great Northern empire) and the less important but still valuable Lake Superior & Mississippi.

Without being able to sell sufficient Northern Pacific bonds and with Jay Cooke & Co. suffering other setbacks in 1872, the Northern Pacific was virtually out of cash, having only itself to blame. It had purchased twice as many locomotives as needed, built scores of miles of useless track, now owned a mostly docked Missouri steamboat—and the list goes on and on.

However, the final blow was triggered a year later by Custer's 1873 report. If, Wall Street asked in effect, Yellowstone Valley construction was a "war zone," what did this mean for Northern Pacific bond holders? It would be like today asking the public to invest in Syria, Iraq, Afghanistan, or the Ukraine. Fearful of the immediate future, at 10:30

[2]Placating his apprehensive wife, Rosser left the survey shortly after it reached the Yellowstone.

A.M. September 18, 1873, Jay Cooke's frightened partners—led by Fahnestock—shut down his company despite its having twice as many assets as liabilities. Within minutes, the Panic of 1873 began, arguably the nation's second worst depression.

Somehow the Northern Pacific continued operations. With track finally at Bismarck and operational economies made by Cass, third-quarter revenues exceeded expenses, in part due to sales of the first parcels of federal land grants. As predicted, Upper Missouri River traffic began being routed via Bismarck, with passengers and materials freighted east over Northern Pacific track. Under Frederick Billings, the Northern Pacific renewed transcontinental construction in 1878. After Henry Villard seized control of the railroad (1881), he aggressively continued to build and the railroad became a transcontinental line, complete with a grand, "final spike" ceremony in September 1883.

* * *

After being briefly discharged, Rosser was rehired and stayed with the Northern Pacific for almost a decade, in the winter of 1877–78 building a then-famous *ice bridge* across the Missouri. However, after relatively easy Plains construction ended, and with Roberts in Brazil (where he died in 1881), Rosser proved professionally unable to meet the Northern Pacific's increasingly difficult construction requirements and was discharged. His employment with the Canadian Pacific ended with his being fired for "inside" investments. Rosser "retired" to Virginia, but never again held steady work. In 1905, President Roosevelt appointed him Charlottesville's postmaster, a position he held, despite a debilitating stroke, until his 1910 death.

Henry C. Davis was rehired by the Northern Pacific. He rose rapidly, took part in the final spike ceremonies, left the railroad, and began a hugely successful Wall Street career. Budding artist Charles "Shorty" Graham made his way to New York, painted stage sets, and became one of *Harper's Weekly* lead illustrators. However, when newspaper technology made it possible to reproduce photographs, his career collapsed. Edward Jordan returned to Portland, Maine. Jordan established a highly respected civil engineering practice that continued under his name for sixty years after his 1935 death. Montgomery Meigs stayed with the Northern Pacific through 1873 when his father assisted in getting him a position with the Corps of Engineers. Professionally respected, he continued working until the 1920s.

On January 6, 1906, almost thirty eastern surveyors and friends held a reunion in Minneapolis. Their purpose was to honor Rosser, who had been felled by a stroke and could no longer speak. Attendees included Davis (who arrived in his own railroad car), Graham, Jordan, and Meigs. Numerous Twin City newspaper accounts survive and indicate their decision to publish a book consisting of "remembrances" and Graham's 1872 illustrations. Although the project never advanced, many of Graham's sketches were forwarded to Rosser, and some appear in this volume. As coincidence would have it, on the Friday of the reunion, Hump, a Minneconjou Lakota war chief who had fought against Baker in '72, passed through Minneapolis (on a Northern Pacific train?) while on a visit to President Roosevelt in Washington. Thus, for one final time, the paths of surveyors and Sitting Bull's followers crossed again.

Appendix A

Daily Log for the 1872 Eastern Yellowstone Survey

Day No.	Date	Miles Marched/Surveyed and Comments
1	July 26	Left Missouri opposite Bismarck, 20 miles, camped Buck Creek, late rain.
2	July 27	5 miles, began 6 A.M., waited for Stanley's column, very warm.
3	July 28	Rained until 8 A.M., 2 miles to north fork of the Cannonball, warm & humid.
4	July 29	15 miles, began 5 A.M.; found good water and grass.
5	July 30	18 miles, 4:45 A.M., crossed Heart River, chilly with a raw wind.
6	July 31	13 miles, 8 A.M., chilly, passed Heart Butte mid-morning.
7	Aug 1	15 miles, 5 A.M., delayed by stream; good grass but little wood.
8	Aug 2	19 miles, crossed Heart River midday.
9	Aug 3	22 miles (17 chronometer), began 5 A.M., level plains, camp on Sandy Creek.
10	Aug 4	15 miles (11 chronometer), began 5:30, crossed south Heart River; Rosser saw two Sioux.
11	Aug 5	15 miles, 5 A.M., Rosser and others to Little Missouri to look for crossing.
12	Aug 6	2 miles; laid in camp awaiting Rosser's return.
13	Aug 7	20 miles, 5 A.M., rough hilly road, camped on branch of Little Missouri, no grass.
14	Aug 8	7 miles, rain, road again very rough & hilly.

Day No.	Date	Miles Marched/Surveyed and Comments
15	Aug 9	10 miles, 5 A.M., reached bluffs at noon, river high, fine "petrifications."
16	Aug 10	Little Missouri too high to cross; many collecting fossils.
17	Aug 11	10 miles, all had crossed Little Missouri by 10 A.M., saw 3 old male buffalo.
18	Aug 12	Pettit died & ceremony in A.M., mileage unclear.
19	Aug 13	Problems with route in rolling badlands, 5½ miles, 25 Indians seen.
20	Aug 14	15 miles, 5 A.M., limited water until 4 P.M., warm.
21	Aug 15	14½ miles, reached O'Fallon Creek 2 P.M. and followed it for 6 miles; crossed creek 5 times.
22	Aug 16	14 miles, attacked by Indians at daybreak, followed O'Fallon Creek all day.
23	Aug 17	Reached Yellowstone valley 10 A.M.; Indians sighted but no attack.
24	Aug 18	8 miles, paralleled Yellowstone & reached Powder River by 10 A.M., began to celebrate when Davis was chased by Indians, later a hotter skirmish.
25	Aug 19	Ran 6½ miles of transit line, very hot.
26	Aug 20	One party working; no mileage stated.
27	Aug 21	Both parties working; Eckelson attacked by Indians, very hot.
28	Aug 22	Return begins, very warm; 1½ hours fighting 100+ Indians in O'Fallon Creek valley.
29	Aug 23	Hot; both parties working; 7 miles mapped; no Indians seen.
30	Aug 24	Hot; limited progress, 3 miles; prepared wagons for return to Fort Rice.
31	Aug 25	In camp all day, Rosser exploring with scouts; word about Baker's fight received.
32	Aug 26	Ran 7 miles of line; left O'Fallon Creek.
33	Aug 27	Ran 8½ miles; camped same location as Aug. 14.
34	Aug 28	Rosser & Jordan ran new route saving some miles; ran 4 miles.
35	Aug 29	Warm; limited water, not enough for all stock.
36	Aug 30	Very warm; ran 5 miles, "line will probably have to be abandoned."
37	Aug 31	Warm; reached Cabin Creek, Indians reported & wagons corralled.

Day No.	Date	Miles Marched/Surveyed and Comments
38	Sept 1	In camp, Rosser exploring with scouts.
39	Sept 2	Rained hard, ran 12 miles; "we have got to abandon the line."
40	Sept 3	Intensely hot; ran line 12 miles.
41	Sept 4	Intensely hot; ran 8½ miles, "rough country will not do for line."
42	Sept 5	Violent hail storm, line went past Indian grave site, near O'Fallon Creek again.
43	Sept 6	Cloudy, cool; false Indian alarm.
44	Sept 7	On Sandy Creek, connected with Cabin Creek line.
45	Sept 8	Marched 22 miles; mule broke loose & injured one man.
46	Sept 9	Cold & rainy; "poor luck" & had to abandon line; Rosser exploring.
47	Sept 10	Good camp site; "It begins to look very dubious for our prospects of getting the line through."
48	Sept 11	"Had to abandon line again . . . almost impassable country."
49	Sept 12	Ran 13½ miles on Cabin Creek, up 5 A.M., wagons camped at 11.
50	Sept 13	Warm; "seems like we'll never get out of this cursed creek."
51	Sept 14	Line abandoned, train marched until 5 P.M., camped on Truman's Creek, good view.
52	Sept 15	Cool; Rosser & scouts rode 60 miles in 13½ hours seeking route.
53	Sept 16	Tried tunnel approach but "found impossible to get grade line over the summit."
54	Sept 17	Windy; same camp, tried another route, but "it was not successful."
55	Sept 18	Same camp, trying another line.
56	Sept 19	Snow squall; moved camp 4 miles, trying a new line.
57	Sept 20	Same camp; ran line 9 miles, spotted scouts from Ft. Rice and wagons.
58	Sept 21	Ran 9 miles near Pettit Creek; Rosser received news of son's death.
59	Sept 22	Met wagons at Little Missouri River that came from Fort Rice with supplies.
60	Sept 23	Rain; no line work; "good many men were full to-day."
61	Sept 24	Rain & cold wind; 5 miles partially in Little Missouri valley.
62	Sept 25	Windy & disagreeable; stayed in valley.

Day No.	Date	Miles Marched/Surveyed and Comments
63	Sept 26	Wagons left for Fort Rice; a surveyor fired for drinking.
64	Sept 27	Warm; left eastern side of valley; Eastman drunk.
65	Sept 28	Cold, heavy rain; good line but only by exceeding technical limits.
66	Sept 29	Warm; Rosser "anxious" to leave, made 9 miles.
67	Sept 30	Warm; Rosser exploring all day, ran only 2 miles.
68	Oct 1	Very warm; Jordan doing paperwork; ran 13½ miles.
69	Oct 2	10 miles; "Eckelson ran Jordan's instrument."
70	Oct 3	Strong P.M. wind, ran 13 miles. [Crosby killed 20 miles away.]
71	Oct 4	Camp moves 19 miles; Gall's attacks kills Steve Harris, Adair mortally wounded.
72	Oct 5	Adair dies 4:40 A.M.; continued the line and ran 7 miles, no Indians.
73	Oct 6	Camp moved to east side of Heart River; Harris buried.
74	Oct 7	Warm; ran 6 miles to Heart River, near Heart Butte, tracks of 100 Indians seen.
75	Oct 8	Windy; ran line 8 miles.
76	Oct 9	Cold, rain, very windy; made only 4 miles, received news of Crosby's death.
77	Oct 10	Warm; ran over 7 miles; Louis Agard left for Ft. Rice.
78	Oct 11	Warm; ran 5¾ miles.
79	Oct 12	Warm; camp on Big Muddy Creek, Stanley about to leave.
80	Oct 13	Warm; some men hunting, surveyors preparing to head northeast.
81	Oct 14	Warm; Stanley & wagons leave; ran 8½ miles toward "the Crossing" [Bismarck].
82	Oct 15	Cool; made 7 miles, Stanley sends reinforcements.
83	Oct 16	Ran 7 miles through considerable brush.
84	Oct 17	Warm; ran 7 miles.
85	Oct 18	Warm; ran 3½ miles, getting close to the Missouri.
86	Oct 19	Warm; party beginning to split up, some at Fort McKean.
87	Oct 20	Warm; most of surveyors at river, Smith drunk.
88	Oct 21	Windy, cold; ran 3 miles.
89	Oct 22	Warm; connected line with Meigs in afternoon; survey over; final entry: "Scout brought up word to Col. Carlin that 250 Indians are coming to attack post."

Source: NPRR *Records, 137.K.13.3 B, Minnesota Historical Society, St. Paul.*

Appendix B

Daily Log for the 1872 Western Yellowstone Survey

Day No.	Date	Miles Marched/Surveyed and Comments
1.	July 27	Infantry, surveyors, and wagon trains leave Fort Ellis to cross Bozeman Pass.
2.	July 28	Cavalry leaves Fort Ellis, crosses pass and catches up with wagons.
3/4.	July 29–30	Camp just south of Shields River. Rope ferry used to reach Crow reservation on east side of Yellowstone, river is too high to cross. No attempt to build larger ferry.
5.	July 31	March along south base of Sheep-Head Mountain on northwestern side of river.
6.	Aug. 1	Reached south flowing Cottonwood Creek.
7.	Aug. 2	7 miles. Ravines make difficult going. Strong current with Yellowstone falling 11 feet per mile.
8.	Aug. 3	5+ miles. Ravines. River 300 feet wide, "booms along in great waves at 7 or 8 miles an hour."
9.	Aug. 4	East of Big Timber. Cross and follow Sweet Grass Creek valley.
10.	Aug. 5	Tributary of creek. At times use ropes to raise/lower wagons over steep crossings.
11.	Aug. 6	Continued up the valley with good water, wood, and grass in abundance.
12.	Aug. 7	8+ mile average past few days. Reached plains dividing Musselshell and Yellowstone.
13.	Aug. 8	4 miles. Bad weather.
14.	Aug 9	Better country with more game. Likely between Rygate and Billings.

Day No.	Date	Miles Marched/Surveyed and Comments
15.	Aug. 10	"After a march of only six miles." Baker drinking? Rolling country.
16.	Aug. 11	18 miles. Unable to get back to Yellowstone. Drier country.
17.	Aug. 12	Reach campsite just east of Billings.
18.	Aug. 13	In camp. Surveyors backtrack 7 miles reaching 1871 survey stakes. Indian signs ignored.
19.	Aug. 14	Indians attack about 3 A.M., but beaten back. Surveyors go 3 miles in midafternoon.
20.	Aug. 15	5+ miles; high anxiety about Indians.
21.	Aug. 16	Camp not moved. Surveyors work nearby. River too high to cross.
22.	Aug. 17	Camp moved across creek down to river. Signs of strong Indian forces.
23.	Aug. 18	Haydon questions Baker on safety. Early A.M. alarms. Camp not moved.
24.	Aug. 19	Camp not moved. Haydon back 6 P.M., "compelled to walk six miles before beginning labors," then 2½ mi, totals 17 in hot, humid days.
25.	Aug. 20	8 miles. Haydon gives up, with request to Baker in writing. Camp again moved forward.
26.	Aug. 21	12 mi., move NW; barren land & only water in buffalo wallows pools. Reach divide.
27.	Aug. 22	Reach wide plateau overlooking Musselshell Valley, much game.
28.	Aug. 23	10 miles, late start allows "nearly famished" animals to graze. Reach Musselshell and see buffalo herds.
29.	Aug. 24	In camp. Men and animals exhausted.
30.	Aug. 25	In camp.
31.	Aug. 26	Surveyors and all cavalry 20 miles to "big bend." All infantry on defensive, in camp.
32.	Aug. 27	Same base camp, surveyors working back west along river.
33.	Aug. 28	Same camp; surveyors now back within a mile.
34.	Aug. 29	Camp moved west; Musselshell swifter and cooler.
35.	Aug. 30	In camp; day "most oppressively warm and dry."
36.	Aug. 31	10 miles; "much skill" shown by Haydon; river passes through "picturesque canyon."
37.	Sept. 1	"Rainy day in camp."
38.	Sept. 2	In camp. During early A.M. thunderstorm, a sentinel fired at a wolf, spooking everyone.

Day No.	Date	Miles Marched/Surveyed and Comments
39.	Sept. 3	Camp not moved; surveyors ran line 3½ miles; weather finally better in afternoon.
40.	Sept. 4	6 miles, difficult going because of river's continuous curves; many fossils, good grazing.
41.	Sept. 5	Violent rain storm; camp not moved; mail from Bozeman (left 8/24) arrives.
42.	Sept. 6	In camp; ground too muddy, makes it impossible to move; much cooler with strong winds.
43.	Sept. 7	Surveyors ran 4½ miles, but ground still too soft for wagons; cold, clear night.
44.	Sept. 8	Wagons 10 miles over hilly ground; heavy thunderstorms.
45.	Sept. 9	6 miles, difficult because of soft ground.
46.	Sept. 10	In camp, with "fires being necessary all day for comfort."
47.	Sept. 11	12 miles, between Judith and Crazy Mountains; valley widens; little wood for fuel.
48.	Sept. 12	In camp.
49.	Sept. 13	14 miles; Musselshell splits.
50.	Sept. 14	In camp. Army carrier with mail and newspapers arrive.
51.	Sept. 15	Handful of miles to South Fork of Musselshell.
52.	Sept. 16	In camp.
53.	Sept. 17	In camp. Surveyors working.
54.	Sept. 18	5 miles. Surveyors working.
55.	Sept. 19	9 miles, on South Fork, in mountains, camp in small level valley, numerous beaver ponds.
56.	Sept. 20	Snow flurries in evening.
57.	Sept. 21	Little timber, some willows and pine; "pure, cold streams wind through the various ravines."
58.	Sept. 22	Drizzle changed into a heavy snow.
59.	Sept. 23	Deep snow prevented change of camp.
60.	Sept. 24	6 miles west, mountains, snow & freezing weather; will march on the Fort Ellis and Camp Baker "road" soon.
61.	Sept. 25	"Camp Separation," area where streams run in four directions; infantry takes "road" north.
62.	Sept. 26	Three cavalry companies with Barlow and Baker go south to Fort Ellis; surveyors and one company head west toward Helena.

Source: John Whitney Barlow, The Report of [Barlow], Who Accompanied a Surveying Party of the Northern Pacific Railroad, in Relation to Indian Interference with That Road, *Senate Committee on the Pacific Railroad, 42nd Cong., 3rd sess., Executive Document no. 16, January 6, 1873.*

Appendix C

Plains Indian Population

This article appears to offer an order-of-magnitude accuracy that is critical in understanding exactly how small the Northern Plains Indian population actually was in 1872.

Sioux City Daily Journal, December 19

INDIANS
THEIR LOCATIONS, TRIBAL NAMES, AND
NUMBERS IN THE DEPARTMENTS
OF THE DAKOTA AND PLATTE

A Chicago reporter has collated, at Gen. Sheridan's headquarters in Chicago, considerable valuable data in regards to the Military Division of Missouri.[1] Within the division are forty-seven Indian agencies and one-hundred and twenty-five tribes or bands upon over sixty reservations, with an aggregate population of 172,205 and a total area of 200,655 square miles in the different reservations, comprising 128,419,200 acres. The Departments of Dakota and Platte are those in which we have local interest, and concerning them we have condensed the following facts and figures.

DEPARTMENT OF DAKOTA

The Department of Dakota includes Minnesota, Montana, and Dakota. In Minnesota there are six reservations [numbering 7,687 Chippewa]. In

[1]Sheridan's Division stretched from Canada to Mexico, and from the Rockies to the Mississippi including Louisiana, Mississippi, Alabama, and western Tennessee.

Montana there are three reservations, one general, the Jocko [far western Montana], and Mountain Crow [upper Yellowstone]. The general reservation includes the Gros Ventre, Piegan, Blood, Blackfeet, Crow, Assiniboines, Northern Cheyenne, Arapahoe, Santee, and Yankton Sioux tribes numbering 23,690; [four Jocko reservations tribes, 1,900]; and Mountain Crow . . . 2,700 strong.

In Dakota there are five reservations. The Devil's Lake reserve includes the Sisseton, Wahpeton, and Cuthead bands of Sioux, numbering 730; Fort Berthold reserve, the Arickera, Gros Ventre, and Mandan tribes, numbering 2,700; General Sioux reserve, Brule, Ogallala, Hunkpapas, Minneconjous, Yanktoni, Blackfeet, Cuthead, Two Kettle, Sans Arc, and Santee bands of Sioux, 28,219 strong;[2] Poncas, 736; Lake Traverse, Sisseton, and Wahpeton bands of Sioux, numbering 1,426.

DEPARTMENT OF THE PLATTE

The Department of the Platte encompasses Iowa, Nebraska, Wyoming, and Utah. In Iowa there are no reservations. In Nebraska there are five reservations: the Winnebago number 1,400; the Omaha, 984; the Otoe and Missouri tribes, 450; the Iowa, the Sac and Fox, 295; and Santee Sioux, 987. In Wyoming the Eastern bands of Shoshone and Bannock number 2,000, and in Utah is one reservation of Utes, numbering 7,800. [Few, if any, of these contributed warriors to Sitting Bull's cause.]

[2]It was from these that Sitting Bull drew the overwhelming majority of his warriors.

Appendix D

Better Off with No Railroad

There is a tendency to assume that railroads in the nineteenth century were economically advantageous, culturally beneficial, and highly regarded. Imagine my surprise, therefore, when reading the *Helena Daily Herald,* that I came across the following letter. At first I assumed that the writer's comments were tongue-in-cheek, but after rereading this, I began to believe that he was serious. Regardless of your interpretation, you will be surprised by the points that he makes (some of which are factually accurate) as well as by the immense changes that the American culture was experiencing.

Helena Daily Herald, November 2

THE OTHER SIDE:
SEVEN IRRESISTIBLY COGENT REASONS
AGAINST RAILROAD
DIRE RESULTS FROM ENCOURAGING THE
ADVANCE OF THE "COWCATCHER OF CIVILIZATION"
LONGEVITY VERSUS THE LOCOMOTIVE

To the Editor of the *Herald*:

As all articles heretofore published in our city's dailies have represented the affirmative side only of the railroad question now being so earnestly discussed by our citizens, permit me to record the views of those who are not only opposed to the subsidizing of the proposed road, but who view with disfavor the construction of any railroad whatever within our Territorial boundaries. Our reasons are embraced in the following briefly stated objections, i.e.:

1st *Railroads increase the perpetration of crime.* In order to prove this, it is but necessary for the minds of "oldsters" to revert to the times when these "broad roads of destruction" were unknown in the land. Then, thieving and burglaries were rare, and murders were never more than a "nine day wonder." *Now*, thieving, in both official and private life, has almost grown respectable, while murders are so frequent that unless the circumstances surrounding their commission are of peculiar atrocity they seldom excite comment and are soon forgotten.

2nd *Railroads aid in the escape of criminals.* In times gone by, when stage coaches and horses only were used in traveling from one section to another, criminals were readily captured and imprisoned, or convicted, as their crimes deserved; but now, crimes are committed with impunity, and the guilty ones flee rapidly by railway conveyance from the scene of their crimes, unapprehended and unpunished.

3rd *Railroads cause domestic unhappiness.* Half a century ago, when railways were first talked about, cases of separation between husband and wife were very unusual, and the party in error was ostracized from respectable society, but now the swift moving cars bring into peaceful towns female adventurers and unprincipled men who prey on domestic felicity, ruining unsophisticated womanhood and luring from their vows of fealty unsuspicious but infatuated men. It is but necessary to cite the reader to the great railways centers, and to call his attention to the unenviable reputation attained by both by reason of the large number of divorces obtained in their courts, to prove the truth of this objection.

4th *Railroads extend the ravages of contagious diseases.* Ills, which might be confined to the locality in which they originated, are made national in their character, by reason of being rapidly conveyed by those infected to far distant sections. As an example of this, we have but to scan our daily telegraphic dispatches and note how very rapidly the horse disease is extending throughout the railway states, creating loss and dismay everywhere.[1]

5th *Railroads shorten the aggregate longevity of man.* We have to but pick up a newspaper and read of the accidents which occur daily to trains in the different sections of our country, by which human beings are killed singly or by the score, aggregating in the year to many thousands, thus

[1]The writer was referring to an undiagnosed disease that swept the country in the fall of 1872, killing millions of horses, especially in larger cities.

suddenly cut off in the midst of their days, when, were there no railroads, they might have lived long years more useful and industrious citizens.

6th *Railroads absorb the capital of the country.* We have but to consider the millions which have been and are being invested in railroads, and the mind stands aghast at the immense amount of capital which is so monopolized. This large sum distributed among our industrial population would greatly increase individual wealth and promote private industry as well as domestic happiness.

7th *Railroads injure pioneer merchants and professional men.* Business men of limited means seek new countries, and by "corners" and "rings" establish a business of profit to themselves. A briefless lawyer, after starving in the old business sections, migrates to a new territory and builds up a lucrative practice. A young doctor (or probably a man without a diploma who has an experimental mind) hangs out his shingle in places remote from easy access, and straightaway patients flock in upon him. A politician vacates his accustomed seat in a "corner grocery," comes West, and is soon elected to a remunerative office and dons good clothes. The construction of a railroad brings others to compete with those having present controls in the trades, professions, and mercantile pursuits and certainly injures if it does not destroy the occupations of the first settlers of the country.

But, Mr. Editor, while there are many other reasons equally as cogent as these given [here], I will refrain from doing so, but will remark that we are of the opinion that any extra session of the Legislature is an unnecessary, as well as an expensive, movement productive of no good but much evil, and if persisted in will lead me and my co-thinkers to migrate to some happy country, where railroads are not and are unknown.

I am proudly,
An Old Fogy

Bibliography

Primary Sources

Anon. Northern Pacific, Dakota Division [A Surveyor's Narrative of 1871–1873]. Albert O. Small Special Collections Library, University of Virginia, Charlottesville; Rosser Collection 1171, Collection 1171-g-h-j, box 12, Miscellaneous Writings.

Baker, Eugene Mortimer. Report of the Operations of the [1872] Escort to Surveyors of the N.P.R.R. Company down the. Retyped copy, Montana Historical Society, Helena.

Barlow, John Whitney. *An Engineer['s] Report of a Reconnaissance of the Yellowstone River in 1871*. U.S. Senate, 42nd Cong., 2nd sess., Executive Document no. 66 (1872).

———. *The Report of [Barlow], Who Accompanied a Surveying Party of the Northern Pacific Railroad, in Relation to Indian Interference with that road.* U.S. Senate, Executive Document No. 16, 42nd Cong., 3rd sess., January 6, 1873, Senate Committee on the Pacific Railroad.

Bennett, Clarence E. August 26, 1872 Letter to Wife. Clarence Edmund Bennett Papers (MS 39) Mansfield Library, University of Montana, Missoula.

Davis, Henry Chandler. Journal: July 26 through September 3, 1872. State Historical Society of Wisconsin, Madison, microfilm 189.

Fahnestock, Harris Charles. Miscellaneous letters to Jay Cooke, October 1868–April 1874. Harvard Business School, Baker Library, F-223, box 3, folder 2.

Irvine, Javan Bradley. Correspondence, diaries, press clippings, obituaries, and miscellaneous, South Dakota State Archives, Cultural Heritage Center, Pierre.

Jordan, Edward C. Account of Edward C. Jordan's Experiences in [the] West during the Construction of the Central Pacific and the Northern Pacific Railroad. Ca. 1906. Compiled and typed by his grand-nephew and namesake, Edward C. Jordan, courtesy of Mark Jordan, Gorham, Me.

Kellogg, Sanford Cobb. "Report to General Sheridan Concerning Stanley's 1872 Yellowstone Expedition." *Bozeman Avant Courier*, November 7, 1872.

Meigs, Montgomery C. "Letters of Montgomery Meigs Written While Engaged in the Survey of the Northern Pacific Railroad, 1872–1873." Edited by Elizabeth Joan

Rodgers Atwater, M.A., State University of Montana, 1937. Microfilm 140, Montana Historical Society, Helena.

Northern Pacific Railroad. Annual Reports, 1869–76. Minnesota Historical Society, St. Paul, microfilm M-369.

———. Clippings and Scrapbooks, 1866–1904, Office of the Secretary of the Northern Pacific. Minnesota Historical Society, microfilm M-522

———. Journal of the Surveying Expedition from the Missouri to the Powder River [1872]. Minnesota Historic Society, 137.k.13 (B).

———. Secretary's Department: Unregistered Letters Received and Related Records, Undated and 1864–1876. Minnesota Historical Society, microfilm M-459.

Ponsford, John W. Statement of the Baker Fight on the Yellowstone River in '72. Burlingame Special Collections Library, Montana State University, Bozeman.

Roberts, Milnor. Correspondence, 1871–1874. Roberts Papers, series 2, box 2, folders 4 and 5.

Rosser, Betty Barbara (Mrs. Thomas L.). 1872 Diary. Albert O. Small Special Collections Library, University of Virginia, Charlottesville, Collection 1171-g-h-j, box 11.

Rosser, Thomas Lafayette. "The Death of Adair." Handwritten and signed, ca. 1895. Albert O. Small Special Collections Library, University of Virginia, Charlottesville, Collection 1171-g-h-j, box 12.

———. Diary for 1872. Albert O. Small Special Collections Library, University of Virginia, Charlottesville, Collection 1171-g-h-j, box 16.

———. Report of Gen. Rosser to D. C. Linsley, Assistant Chief Engineer, October 10, 1872. Minnesota State Historic Society, 134.L13.8(F).

Rumbough, Sarah Eliza Stanley. "The Regular Army O!" Daughters of the U.S. Army [Club], West-Stanley-Wright Family Papers, U.S. Army Military History Institute, Carlisle.

Stanley, David Sloan. Report on the Yellowstone Expedition of 1872. Montana Historical Society, 917.86 Sta. Also *Westerners Brand Book (New York Posse)* 3, nos. 2 and 3 (1956).

U.S. Department of the Interior. Annual Report of the Commissioner of [the U.S. Office of] Indian Affairs for the Year. Washington: Government Printing Office, 1867–74.

U.S. Department of War. Annual Reports. Washington: Government Printing Office, 1866–74.

———. *Official Army Register for January*. Washington: Government Printing Office, 1869–74.

———. Department of the Dakota. *Outline Descriptions [and Detailed Maps] of the Posts in the Military Division of the Missouri Commanded by Lt. Gen. P. H. Sheridan*. Chicago, 1876.

Secondary Sources

American National Biography. 22 vols. New York: Oxford University Press, 1999.

Blackburn, William M. (and notes by De Lorme M. Robinson). *History of North and South Dakota*. Vol. 1. Pierre: South Dakota Historical Collections, 1902.

Blackstead, David J., ed. *Bismarck [at] 100: Official Publication of the Bismarck Centennial Association*. Bismarck: Bismarck Centennial Association, 1972.

Bradley, James H. *The March of the Montana Column: A Prelude to the Custer Disaster*. Edited by Edger I. Stewart. Norman: University of Oklahoma Press, 1991.

Bray, Kingsley M. *Crazy Horse: A Lakota Life*. Norman: University of Oklahoma Press, 2006.

Brown, John K. *The Baldwin Locomotive Works, 1831–1915*. Baltimore: Johns Hopkins University Press, 2001.

Brown, Mark H. *The Plainsmen of the Yellowstone*. New York: G. P. Putnam's Sons, 1961.

Burgum, Jessamine Slaughter. *Zezula or Pioneer Days in the Smoky Water Country: A Collection of Historical Sketches . . . of the Early and Romantic History of [the] Dakota Territory along the Missouri River*. Valley City, N.D.: Getchell & Nielsen, ca. 1930s.

Bursack, William W., and Jerry Masters. "The Northern Pacific before the Last Spike: Locating the Route." Unpublished, 2004.

Bushong, Millard K., and Dean M. *Fightin' Tom Rosser, C.S.A.* Shippensburg, Pa.: Beidel Printing House, 1983.

Chittenden, Hiram Martin. *History of Early Steamboat Navigation on the Missouri River: [The] Life and Adventures of Joseph La Barge, Pioneer Navigator and Indian Trader for Fifty Years Identified with the Commerce of the Missouri Valley*. 2 vols. New York: Francis P. Harper, 1903.

———. *The Yellowstone National Park*. Edited by Richard A. Bartlett. 1895; reprint, Norman: University of Oklahoma Press, 1964.

Cullum, George W. *Biographical Register of the Officers and Graduates of the United States Military Academy*. Boston: Houghton Mifflin, 1891.

Eckroth, David. *Sitting Bull's Fight with Colonel Stanley, August 22, 1872, "The Battle of O'Fallon Creek."* Privately printed for the American Battlefield Protection Program (Program Grant # GA-2255-08-0050), 2010.

Ege, Robert J. *"Tell Baker to Strike Them Hard": Incident on the Marias, 23 Jan. 1870*. Bellevue, Neb.: Old Army Press, 1970.

Frey, Robert Lovell, ed. *Encyclopedia of American Business History and Biography: Railroads in the Nineteenth Century*. New York: Facts on File, 1998.

Gray, John S. *Custer's Last Campaign: Mitch Boyer and the Little Big Horn Reconstructed*. Lincoln: University of Nebraska Press, 1993.

Hanson, Joseph Mills. *The Conquest of the Missouri: Being the Story of the Life and Exploits of Captain Grant Marsh*. 1916; reprint, Mechanicsburg, Pa.: Stackpole Books, 2003.

Heitman, Francis B. *Historical Register and Dictionary of the United States Army . . . , from September 29, 1789, to March 2, 1903*. 1903; reprint, Urbana: University of Illinois Press, 1965.

Hobson, Archie, ed. *The Cambridge Gazetteer of the United States and Canada: A Dictionary of Places.* New York: Cambridge University Press, 1995.

Hodge, Frederick W., ed. *Handbook of American Indians North of Mexico*, 2 vols. 1906; reprint, New York: Pageant Books, 1959.

Howard, Joseph Kinsey. *Strange Empire: A Narrative of the Northwest.* 1952; reprint, St. Paul: Minnesota Historical Society, 1994.

Houston, E. Lina. *Early History of Gallatin County.* Bozeman, Mt.: privately printed, 1933.

Hoxie, Frederick E. *Parading through History: The Making of the Crow Nation in America, 1805–1935.* Cambridge: Cambridge University Press, 1995.

Hutton, Paul Andrew. *Phil Sheridan and His Army.* Lincoln: University of Nebraska Press, 1989.

Lamar, Howard R. *The New Encyclopedia of the American West.* New Haven: Yale University Press, 1998.

Larson, Robert W. *Gall, Lakota War Chief.* Norman: University of Oklahoma Press, 2007.

Lass, William E. *A History of Steamboating on the Upper Missouri River.* Lincoln: University of Nebraska Press, 1962.

———. *Navigating the Missouri: Steamboating on Nature's Highway, 1819–1935.* Norman: Arthur H. Clark, 2008.

Lounsberry, Clement A. *Early History of North Dakota.* 1919; reprint, Ann Arbor: University Microfilms International, 1978.

Lubetkin, M. John, ed. *Custer and the 1873 Yellowstone Survey: A Documentary History.* Norman: Arthur H. Clark, 2013.

———. *Jay Cooke's Gamble: The Northern Pacific, the Sioux, and the Panic of 1873.* Norman: University of Oklahoma Press, 2006.

———. *Union College's Class of 1868: The Unique Experiences of Some "Average" Americans.* Privately published, 1995.

Malone, Dumas, ed. *Dictionary of American Biography.* 24 vols. Includes 2 supplements. New York: Charles Scribner's Sons, 1928–44.

Middleton, William D., George M. Smerk, and Roberta L. Diehl, eds. *Encyclopedia of North American Railroads.* Bloomington: Indiana University Press, 2007.

Milner, Clyde A., II, and Carol A. O'Connor. *As Big as the West: The Pioneer Life of Granville Stuart.* New York: Oxford University Press, 2009.

The National Cyclopedia of American Biography. 61 vols. New York: James T. White, 1898–1982.

Oberholtzer, Ellis Paxson. *Jay Cooke, Financier of the Civil War.* 2 vols. 1907; reprint, New York: Augustus M. Kelley, 1968.

Perret, Geoffrey. *Ulysses S. Grant: Soldier & President.* New York: Random House, 1997.

Powell, Peter John. *People of the Sacred Mountain: A History of the Northern Cheyenne Chiefs and Warrior Societies, 1830–1879.* 2 vols. San Francisco: Harper and Row, 1981.

Powell, William H. *List of Officers of the Army of the United States from 1799 to 1900.* 1900; reprint, Detroit: Gale Research Company, 1967.

Renz, Louis Tuck. *The Construction of the Northern Pacific Railroad.* Walla Walla, Wash.: privately printed, 1973.

———. *The History of the Northern Pacific Railroad.* Fairfield, Wash.: Ye Galleon Press, 1980.

Robinson, Doane. *A History of the Dakota or Sioux Indians.* 1904; reprint, Minneapolis: Ross & Haines, 1956.

Sandburg, Carl. *Abraham Lincoln: The War Years.* Vol. 2. New York: Harcourt, Brace, 1939.

Sandoz, Mari. *Crazy Horse: The Strange Man of the Ogalas.* 1942; reprint, Lincoln: University of Nebraska Press, 1992.

Schrenk, Lorenz P., and Robert L. Frey. *The Northern Pacific Pioneer Steam Era.* Minneapolis: Monad Press, 2010.

Sharp, Paul F. *Whoop-Up Country: The Canadian-American West, 1865–1885.* 1955; reprint, Norman: University of Oklahoma Press, 1973.

Slaughter, Linda W. *From Fortress to Farm; or, Twenty-Three Years on the Frontier.* Edited by Helen Eastman. 1895; reprint, New York: Exposition Press, 1972.

Smalley, Eugene V. *History of the Northern Pacific Railroad.* 1883; reprint, New York: Arno Press, 1975.

Smith, Phyllis. *Bozeman and the Gallatin Valley: A History.* Helena, Mont.: Falcon Press, 1996.

Stearns, Harold Joseph. *A History of the Upper Musselshell Valley of Montana [to 1920].* Harlowton, Mont.: Times-Clarion, 1966.

Stewart, Edgar I. *Custer's Luck.* 1955; reprint, Norman: University of Oklahoma Press, 1980.

Taft, Robert. *Artists and Illustrators of the Old West, 1850–1900.* 1953; reprint, Princeton: Princeton University Press, 1982.

Tate, Michael L. *The Frontier Army and the Settlement of the West.* Norman: University of Oklahoma Press, 1999.

Taylor, Joseph H. *Sketches of Frontier and Indian Life on the Upper Missouri.* Bismarck, N.D.: privately published, 1897. Also known as "Frontier and Indian Life."

———. *Kaleidoscopic Lives: A Companion Book to "Frontier and Indian Life."* Washburn, N.D.: privately published, 1902.

Thrapp, Dan L. *Encyclopedia of Frontier Biography.* Lincoln: University of Nebraska Press, 1991.

Topping, E. S. *Chronicles of the Yellowstone.* 1884; reprint, Minneapolis: Ross & Haines, 1968.

U.S. Works Projects Administration [WPA], Federal Writers Project. *Minnesota: A State Guide.* 1938; reprint, New York: Hastings House, 1954.

———. *WPA Guide to 1930s Montana.* 1939; reprint, Tucson: University of Arizona Press, 1994.

———. *WPA Guide to 1930s North Dakota.* 1938; reprint, Bismarck: State Historical Society of North Dakota, 1990.

Utley, Robert M. *Frontier Regulars: The United States Army and the Indian, 1866–1891.* New York: Macmillan, 1973.

———. *The Indian Frontier and the American West.* 1983; reprint, Albuquerque: University of New Mexico Press, 1993.

———. *The Lance and the Shield: The Life and Times of Sitting Bull.* New York: Henry Holt, 1993.

Veenendaal, Augustus J., Jr. *The Saint Paul & Pacific Railroad: An Empire in the Making.* DeKalb: Northern Illinois University Press, 1999.

Vestal, Stanley. *Sitting Bull: Champion of the Sioux.* 1932; reprint, Norman: University of Oklahoma Press, 1989.

Warner, Ezra J. *Generals in Blue: Lives of the Union Commanders.* Baton Rouge: Louisiana State University Press, 1964.

Wert, Jeffry D. *Custer: The Controversial Life of George Armstrong Custer.* New York: Simon & Schuster, 1996.

White, John H., Jr. *A History of the American Locomotive: Its Development, 1830–1880.* New York: Dover, 1980.

Winser, Henry J. *The Great Northwest: A Guide-Book and Itinerary for the Use of Tourists and Travelers over the Lines of the Northern Pacific Railroad.* New York: G. P. Putnam's Sons, 1883.

Articles, Brochures, Essays, Monographs, and Presentations/Speeches

Adler, Bertram. "Old Timer Tales—No. 4: Surveying the Bad Lands." *Railroad Man's Magazine*, February (late 1920s). Reprinted article in North Dakota State Historical Society archives.

Bray, Kingsley M. "Teton Sioux Population History, 1655–1881." *Nebraska History*, Summer 1994.

DeLand, Charles E. "Basil Clement (Claymore): The Mountain Trappers." *South Dakota Historical Collections*, 1922.

Freedom, Gary S. "Moving Men and Supplies: Military Transportation on the Northern Great Plains, 1866–1891." *South Dakota History* 14, no. 2 (Summer 1984).

Grinnell, George Bird. "Recollections of Charley Reynolds." *Westerners Brand Book (New York Posse)* 3, no. 3 (1956).

Hammer, Kenneth M. "Railroads and the Frontier Garrisons of Dakota Territory." *North Dakota History* 46, no. 3 (Summer 1979).

———. "Territorial Towns and the Railroads." *North Dakota History* 36, no. 4 (Fall 1969): 357–69.

Hanson, Joseph Mills. "Thomas Lafayette Rosser." *Cavalry Journal*, March–April 1934.

Harnsberger. John L. "Jay Cooke, the Northern Pacific [Railroad], and the Founding of Bismarck, D.T." *North Dakota Quarterly*, Summer 1981.

———. "Railroads to the Northern Plains, 1870–1872." *North Dakota Quarterly*, Summer 1959.

Harnsberger. John L., and Robert P. Wilkins. "Transportation on the Northern Plains: I. The Genesis of Commerce." *North Dakota Quarterly*, Winter 1961.

———. "Transportation on the Northern Plains: III. The Railroads Arrive." *North Dakota Quarterly*, Summer 1961.

Hutton, Paul Andrew. "Phil Sheridan's Pyrrhic Victory: The Piegan Massacre, Army Politics, and the Transfer Debate." *Montana*, Spring 1982.

Johnson, Roy P. "Maj. Gen. Thomas L. Rosser." A series of articles originally published by the Fargo *Forum* in 1950 and reprinted on April 2, 9, 16, 23 and 30, and May 7, 14, 21 and 28, 1978.

Lass, William E. "Steamboats on the Yellowstone." *Montana*, Autumn 1985.

Lubetkin, M. John. "An Ambush That Changed History: Thomas Rosser, David Stanley, Gall, and the 1872 Eastern Yellowstone Survey." *Journal of the West* 43, no. 1 (Winter 2004).

———. "The 1906 [Surveyors'] Reunion: Thomas L. Rosser and the Northern Pacific Pioneers of 1873." *Mainstreeter* 22, no. 2 (Summer 2003).

———. "'No Fighting Is to Be Apprehended': Major Eugene Baker, Sitting Bull, and the Northern Pacific Railroad's 1872 Western Yellowstone Surveying Expedition." *Montana* 56, no. 2 (Summer 2006).

Mangum, Neil. "Gall: Sioux Gladiator or White Man's Pawn." Fifth Annual Symposium, Custer Battlefield & Historical Museum Association, Hardin, Mt., 1991.

Mattison, Ray H. "Fort Rice: North Dakota's First Missouri River Military Post." *North Dakota History* 20 (1953).

McClernand, Lt. Edward J. *On Time for Disaster: The Rescue of Custer's Command.* 1926; reprint, Lincoln: University of Nebraska Press, 1989.

Parsons, John E. "Sioux Resistance on the Yellowstone, 1872." *Westerners Brand Book (New York Posse)* 3, no. 1 (1956).

Robertson, Francis B. "'We Are Going to Have a Big Sioux War': Colonel David S. Stanley's Yellowstone Expedition of 1972." *Montana* 34, no. 4 (Autumn 1984).

Sandoz, Mari. "Some Oddities of the American Indian." *Westerners Brand Book (Denver Posse)*, 1955, 17–28.

Sheldon, Robert C. "Surveying the Dakotas." *North Dakota Quarterly*, Winter 1961.

Sibbald, John R. "Frontier Inebriates with Epaulets." *Montana*, Summer 1969.

Taylor, Joseph Henry. "Bloody Knife and Gall." *North Dakota Historical Quarterly* 4, no. 3 (April 1930): 165–73.

———. "Lonesome Charley." *North Dakota Historical Quarterly* 4, no. 4 (July 1930): 227–38.

U.S. Military Academy. *Annual Reunion of the Association of the Graduates of the United States Military Academy.* West Point, N.Y., various issues.

Vestal, Stanley. "White Bull and One Bull: An Appreciation." *Westerners Brand Book (Chicago Posse)* 4, no. 8 (October 1947).

Wilson, Wesley C. "The U.S. Army and the Piegans: The Baker Massacre on the Marias." *North Dakota History* 32, no. 1 (January 1965).

Newspapers and Periodicals

Army and Navy Journal, weekly, May 1869–July 1874.
Bozeman Avant Courier, weekly, September 1871–December 1873.
Brainerd Tribune, weekly, February 1872–September 1873.
Chicago Inter-Ocean, daily, May–October 1872, June–September 1873.
Commercial and Financial Chronicle and Hunt's Merchants' Magazine, New York, weekly, July 1871–December 1873.
Duluth Minnesotian, weekly, January 1870–December 1871.
Frank Leslie's Illustrated Newspaper, weekly, September 1870–February 1874.
Harper's New Monthly Magazine, June 1870–May 1874.
Harper's Weekly, January 1871–December 1873.
Helena Daily [*and Weekly*] *Herald*, January 1869–November 1873.
Mainstreeter, quarterly, 1995 to present, Northern Pacific Railway Historical Association, Kirkland, Wash.
Minneapolis Tribune, daily, selected dates 1871–73.
New York Herald, daily, June–October 1872, May–October 1873.
New York Times, daily, spring 1869–winter 1873.
New York Tribune, daily, July–October 1872, June–October 1873.
Northwest Magazine, monthly, January 1883–December 1889.
Philadelphia Evening Bulletin, daily, August–October 1872, August–September 1873.
Philadelphia Inquirer, daily, July–December 1872, July–October 1873.
Railroad Gazette, weekly, January 1870–December 1873.
Red River Star, Moorhead, weekly, July 1872–December 1873.
St. Paul Pioneer, daily and/or weekly, April 1871–October 1873.
St. Paul Press, daily and/or weekly, 1869–73.
Sioux City Journal, daily, June–December 1871 and July 1872–September 1873.
Yankton [*S.D.*] *Herald*, weekly, April 1872–December 1872.

Acknowledgments

This book was essentially completed in the winter of 2008–2009. Then came the big, bad Recession. If I had done a lot of work on this, so had Bob Clark, then editor of the Arthur H. Clark Company, a University of Oklahoma Press imprint. Without his herculean efforts I don't think *Before Custer* would ever have seen the light of day. Since Bob's return to the less tornado-challenged Northwest, editor-in-chief Chuck Rankin and managing editor Steven Baker of the University of Oklahoma Press have seen this through, and hats off to both.

David Eckroth is the co-discoverer of both the Baker and O'Fallon Creek battle sites. Dave has shared his information with me, and my pictures at both locations were taken when he showed me around. Without his time and generosity, this book simply wouldn't have worked.

For this and *Jay Cooke's Gamble*, I received help from almost two hundred people in various libraries, especially the Fairfax and Arlington County, Virginia, public library systems, as well as the Alderman Special Collections Library, University of Virginia–Charlottesville; Arlington County; Burlingame Special Collections Library, Montana State University–Bozeman; Harvard University's Houghton Reading Room, Cambridge; the Library of Congress and National Archives in Washington, D.C., and Silver Spring, Md.; Minnesota State Historic Society, St. Paul; Montana State Historic Society, Helena; North Dakota State Historic Society, Bismarck; South Dakota State Historic Society, Pierre; U.S. Army Military Academy, Archives and Special Collections Department, West Point; and the U.S. Army Military History Institute, Carlisle, Pa.

Invaluable support has also been provided by many members of the Custer Battlefield Historical and Museum Association, the Little Big Horn Associates, and the Northern Pacific Railway Historical Association. Each organization holds annual conferences, publishes high-quality magazines, and involves intelligent, well-informed, and friendly individuals who have gone out of their way to be helpful.

As a reader often frustrated by good books with bad maps—if any—I have tried to use period or easy-to-read maps. For the latter I am particularly appreciative of the work of Vicki Trego Hill, El Paso, Texas, who took period maps and my rough sketches and turned them into something first-rate.

My son, William Andrew Lubetkin, has painstakingly brought me into the electronic wonders of the 1990s—perhaps even the twenty-first century. And I especially wish to thank my wife, Linda, a retired high school English teacher, who spent hundreds of hours reading and making numerous critical suggestions and comments. Without her skills, patience, and good humor, this book couldn't have been completed.

Index

Page references in italic type refer to illustrations.